Springer Theses

Recognizing Outstanding Ph.D. Research

For further volumes:
http://www.springer.com/series/8790

Aims and Scope

The series "Springer Theses" brings together a selection of the very best Ph.D. theses from around the world and across the physical sciences. Nominated and endorsed by two recognized specialists, each published volume has been selected for its scientific excellence and the high impact of its contents for the pertinent field of research. For greater accessibility to non-specialists, the published versions include an extended introduction, as well as a foreword by the student's supervisor explaining the special relevance of the work for the field. As a whole, the series will provide a valuable resource both for newcomers to the research fields described, and for other scientists seeking detailed background information on special questions. Finally, it provides an accredited documentation of the valuable contributions made by today's younger generation of scientists.

Theses are accepted into the series by invited nomination only and must fulfill all of the following criteria

- They must be written in good English.
- The topic should fall within the confines of Chemistry, Physics, Earth Sciences, Engineering and related interdisciplinary fields such as Materials, Nanoscience, Chemical Engineering, Complex Systems and Biophysics.
- The work reported in the thesis must represent a significant scientific advance.
- If the thesis includes previously published material, permission to reproduce this must be gained from the respective copyright holder.
- They must have been examined and passed during the 12 months prior to nomination.
- Each thesis should include a foreword by the supervisor outlining the significance of its content.
- The theses should have a clearly defined structure including an introduction accessible to scientists not expert in that particular field.

Thomas Farmer

Structural Studies of Liquids and Glasses Using Aerodynamic Levitation

Doctoral Thesis accepted by
the University of Bristol, UK

Springer

Author
Dr. Thomas Farmer
H.H. Wills Physics Laboratory
School of Physics
University of Bristol
Bristol
UK

Supervisor
Dr. Adrian Barnes
H.H. Wills Physics Laboratory
School of Physics
University of Bristol
Bristol
UK

ISSN 2190-5053 ISSN 2190-5061 (electronic)
ISBN 978-3-319-36054-6 ISBN 978-3-319-06575-5 (eBook)
DOI 10.1007/978-3-319-06575-5
Springer Cham Heidelberg New York Dordrecht London

© Springer International Publishing Switzerland 2015
Softcover reprint of the hardcover 1st edition 2015
This work is subject to copyright. All rights are reserved by the Publisher, whether the whole or part of the material is concerned, specifically the rights of translation, reprinting, reuse of illustrations, recitation, broadcasting, reproduction on microfilms or in any other physical way, and transmission or information storage and retrieval, electronic adaptation, computer software, or by similar or dissimilar methodology now known or hereafter developed. Exempted from this legal reservation are brief excerpts in connection with reviews or scholarly analysis or material supplied specifically for the purpose of being entered and executed on a computer system, for exclusive use by the purchaser of the work. Duplication of this publication or parts thereof is permitted only under the provisions of the Copyright Law of the Publisher's location, in its current version, and permission for use must always be obtained from Springer. Permissions for use may be obtained through RightsLink at the Copyright Clearance Center. Violations are liable to prosecution under the respective Copyright Law.
The use of general descriptive names, registered names, trademarks, service marks, etc. in this publication does not imply, even in the absence of a specific statement, that such names are exempt from the relevant protective laws and regulations and therefore free for general use.
While the advice and information in this book are believed to be true and accurate at the date of publication, neither the authors nor the editors nor the publisher can accept any legal responsibility for any errors or omissions that may be made. The publisher makes no warranty, express or implied, with respect to the material contained herein.

Printed on acid-free paper

Springer is part of Springer Science+Business Media (www.springer.com)

To Anne and Patrick Handy, in celebration of 60 years of marriage

Supervisor's Foreword

Glass, a disordered form of condensed matter, has been valued technologically for centuries. Transparent oxide glasses were developed for and continue to be valued in the construction industry as they allow light to penetrate easily into large buildings, while being resistant to corrosion from the atmosphere. This resistance is also key to their other wide ranging uses, such as in chemical processing and analysis, and household tableware. More recently, the development of high grade optical fibers has driven the high speed Internet revolution, and the doping of similar glasses has given rise to such innovations as high power fiber lasers.

Despite these extensive developments and uses, the formation of glasses through what is termed the glass transition, remains very poorly understood, and a universal theoretical description of the glass forming process remains elusive. As a result, even the most elementary aspects of glass science are not fully understood, such as why some materials easily form glass whereas others always crystallize, and the bonding and coordination of different atoms.

In crystalline materials the periodic arrangement of the atoms in repeating unit cells allows an accurate knowledge of the atomic arrangements to be ascertained by diffraction methods, these were established in the early twentieth century and are now routinely applied using X-rays, neutrons, and electrons as the probe. In contrast to crystalline materials, where diffraction is characterised by Bragg peaks whose intensity and position is directly related to the crystal structure, the diffraction pattern from a glass reveals no more than a series of diffuse diffraction rings. These rings are related to the statistical arrangement of all of the atoms in the diffraction volume defined by the correlations between the atoms in terms of the radial distribution function. The problem for glasses containing many atom types is that the single diffraction pattern obtained is a mixture of all the correlations between different atom types. As a consequence the individual correlations cannot be isolated by any direct method.

In this thesis Tom Farmer used several recent experimental and analytical developments to explore the properties of two glasses: Barium Titanate glasses and Ytrria–Alumina glasses.

Titanate glasses are difficult to produce by conventional glass forming methods without the addition of a good glass former, for example SiO_2, to improve their glass forming ability. In this work Tom used aerodynamic levitation and laser heating to produce titanate glasses without the need for such additions. The key to the technique is that small beads (of the order of 2–3 mm diameter) of the target material can be levitated and heated (by direct laser heating) and then cooled under containerless conditions. The key advantages are that there is no contamination from a container when the material is heated, and that heterogeneous crystal nucleation is avoided when cooling under containerless conditions, such that the liquid can be highly supercooled enhancing the likelihood of glass formation. Tom applied neutron diffraction, X-ray absorption methods and molecular dynamics and Monte Carlo methods to obtain detailed structural information on the atomic scale for these glasses. A clear observation is that, unlike most good oxide glass formers, the Ti in these glasses adopts a variety of coordination environments (four-fold tetrahedral coordination is typical in for example Si, Al and Ge-based oxide glasses). Furthermore, it is found that these glasses allow a high degree of rare-earth doping and Tom was also able to establish the structure around these atoms—a property that is crucial for optimizing their optical activity.

Yttria-aluminate liquids and glasses have received much attention and provoked controversy in recent years, as it is reported that the liquid can undergo a liquid–liquid transition without phase separation in the supercooled liquid at ~1700 K. As this temperature is deep into the supercooled region of the liquid (it melts at ~2200 K) it is only accessible using aerodynamic levitation and laser heating techniques. In a challenging experiment involving the mounting of this system at the small angle neutron scattering instrument D22 at the ILL, Tom was able to demonstrate that there was no evidence of a phase transition at this temperature and that the observed effects were most likely due to formation of nanocrystals in the liquid on cooling.

In addition to these main themes, Tom also demonstrated how the aerodynamic levitation method could be successfully applied to wide angle neutron diffraction experiments as applied to the chemical and magnetic ordering in liquid Fe–Ni invar alloys at high temperatures.

Bristol, April 2014 Dr. Adrian Barnes

Abstract

In this thesis aerodynamic levitation and laser heating has been combined with neutron diffraction, X-ray absorption spectroscopy (XAS), and computer simulations to investigate the structure of liquids and glasses.

In order to determine the structure of $BaTi_2O_5$ glass and $RE_{0.3}Ba_{0.7}Ti_2O_{5.15}$ glass the complementary techniques of neutron diffraction and XAS were combined. Molecular dynamics models based on simple empirical potentials were generated and then refined using reverse monte carlo (RMC) simulations fitting to the neutron diffraction data. From these structural models the glasses were determined to consist of a network of 4, 5, and 6 coordinated Ti–O polyhedra, which was found to be consistent with the X-ray absorption near edge spectra. A decrease in the rare earth-oxygen bond length with increasing rare earth atomic mass was observed in both the RMC models and the rare earth edge XAS.

The contentious issue of the apparent iso-compositional liquid–liquid phase transition in the $(Y_2O_3)_x(Al_2O_3)_{1-x}$ system was investigated using in situ aerodynamic levitation and small angle neutron scattering (SANS). Samples of $x = 0.2$, 0.25, 0.3, and 0.375 were studied across a temperature range of 1300–1900 K, with detailed measurements at the reported liquid–liquid transition temperature of 1788 K [1]. There was no observed increase in SANS intensity consistent with the nucleation of a second liquid phase at any temperature.

In situ aerodynamic levitation and neutron diffraction with isotopic substitution was used to determine the atomic structure of liquid Invar ($Fe_{65}Ni_{35}$) at ~1800 K. Although the introduction of Ni is substitutional, a small degree of chemical was apparent in the Bhatia–Thornton (1970) partial structure factors. Significant magnetic correlations were detected ~1300 K above the Invar Curie temperature.

Anomalous neutron scattering was undertaken on the unusual semiconducting liquid InSe. A first peak coordination number of 3 suggested that the underlying cause of the unusual behavior is not related to In–In homopolar bonding.

Reference

1. Greaves GN, Wilding MC, Fearn S, Langstaff D, Kargl F, Cox S, Van Q Vu, Majérus O, Benmore CJ, Weber R, Martin CM, Hennet L (2008) Detection of first-order liquid/liquid phase transitions in yttrium oxide–aluminum oxide melts. Science 322:566–570

Acknowledgments

I would like to thank my supervisor Dr. Adrian Barnes for his expertise, advice, and motivation throughout my postgraduate studies.

I am very grateful to Gavin Alexander, Dr. Robert Greasty, Dr. Pinit Kidkhunthod and Laura Mears for many useful discussions, their encouragement, and generally putting up with me. Thanks also to Dr. Henry Fischer, Dr. Louis Hennet, Dr. Anita Zeidler, Prof. Phil Salmon, Dr. Guillaume Ferlat, and Dr. Chris Jones for their contributions to this work.

Finally, I would like to thank my parents, my sister, and my fiancée, whose support and guidance has allowed me to undertake this work.

Contents

Chapter 1
Introduction

The study of the structure of liquids and glasses is of both great fundamental and technological importance [21, 31]. While it may be more commonly associated with determining crystal structures, neutron diffraction is also an important probe of the short and intermediate range ordering in amorphous materials [16, 31, 41]. As with crystals, the two different interaction mechanisms of neutrons with matter allows investigation into both the atomic and magnetic structure. Combined with this is fact that the interaction strength varies significantly depending on the isotope, the classic example being the difference in scattering lengths of hydrogen (−3.7409 fm) and deuterium (6.674 fm) [13]. For samples of more than one element, this enables the atomic positioning of each element to be determined. This is achieved by combining diffraction patterns from samples with identical compositions but different isotopes, which is called isotopic substitution [14].

Neutron diffraction of amorphous materials has been utilized for a very wide range of fields, from the homopolar bonding in glassy $GeSe_2$ [34] to the investigation of planetary core liquids [10], with new techniques continually accessing previously inaccessible areas of phase diagrams.

The relatively recent development of contactless processing, such as aerodynamic [32], aero-acoustic [42], and electrostatic [38] levitation techniques has enabled the structural determination of amorphous materials that were previously experimentally challenging or even unachievable. Levitation methods have been used for both in situ structural measurements [9, 23] and ex situ for atypical sample preparation; this includes sample purification [43, 47] and single crystal growth [6], but is perhaps most often applied to the fabrication of new glasses which are unobtainable by other methods.

The quenching of a liquid to a glass is dependent on avoiding both heterogeneous and homogeneous nucleation [25]. Assuming complete sample purity (i.e. no contaminant seeding), heterogeneous nucleation can only occur at a solid interface, which is inherently avoided by contactless processing. When combined with fast quenching (~400 K/s), such as that achieved with laser heating, it becomes possible to significantly increase the compositional range of glass forming regions, making it the ideal

T. Farmer, *Structural Studies of Liquids and Glasses Using Aerodynamic Levitation*, Springer Theses, DOI: 10.1007/978-3-319-06575-5_1,

© Springer International Publishing Switzerland 2015

tool for glass fabrication. Another notable advantage to this technique is the ability to precisely control the rate at which a sample is quenched. While the maximum quench rate is obviously achieved by instantaneous laser shutoff, a continual reduction in power allows essentially any lower quench rate to be selected. An interesting example of the distinct change in $BaTiAl_2O_6$ caused by varying the quench rate is given by Skinner et al. [39].

Determining the atomic structure of these glasses is important for two reasons. From a technological perspective a detailed structural understanding will allow for a greater degree of tunability of the glass properties [17]. When compared with both the liquid structure and the corresponding crystal structure, the atomic configuration of glasses may reveal the conditions under which a glass will form. Ultimately when combining this information with kinetic measurements this could result in a complete picture of the glass transition, which is currently lacking [3, 4].

The glass formation in $BaTi_2O_5$ (BTO) and rare earth doped BTO is of interest from both a technological and theoretical perspective. Due to significant ferroelectric properties of the underlying crystal [2], the controlled quenching of BTO may allow for the production of glass ceramics exhibiting a strong ferroelectric effect [49]. BTO glass may also be a good candidate for doping with fluorescent rare earth ions, as its low phonon energy should reduce the decay of fluorescence compared with rare earth doped aluminosilicate glass [26, 45]. An important factor in the multiphonon quenching of the fluorescence is the clustering of the rare earth ions [5]. In this work a combination of neutron diffraction, X-ray absorption spectroscopy (XAS), and Reverse Monte Carlo (RMC) computer modelling will be used to investigate the atomic structure, including the proportion of neighbouring rare earth atoms.

Aside from focusing specifically on glass formation, contactless processing is also ideally suited to studying high melting point liquids, which are otherwise prone to container contamination. Two diverse applications of the importance of this capability are demonstrated in this thesis; first it is applied to the study of a reported first order iso-composition liquid–liquid phase transition in the yttria aluminate system [18].

The intriguing proposition of a first order liquid–liquid phase transition between a high density, high temperature liquid and a low density, low temperature liquid was initially made by Rapaport [36, 37] in order to explain melting curve maxima. Although the concept of coexistence between two liquids of identical compositions but different densities appears to be counter-intuitive, indications of it has been reported in several systems, including water [27], phosphorus [22], and Triphenyl phosphate [8]. However in all of these systems there is a significant amount of debate about whether the observations can be attributed to a liquid–liquid phase transition [11, 12, 20, 28].

This is also a contentious issue in the yttria aluminates [1, 7, 29, 30, 40], and has been investigated in both the resulting amorphous state [1, 29] and using simulations [46], but has only recently been studied in situ as the transition occurs in the supercooled state [7, 18]. In this thesis, in situ small angle neutron scattering and pyrometric studies were undertaken in an attempt to reduce the uncertainty about this potential new phenomenon.

The second application is to the long standing Invar problem [35], which is the anomalously low coefficient of thermal expansion of some crystalline metal alloys, most notably $Fe_{65}Ni_{35}$. The Invar (for invariant) effect was discovered by Guillaume [19], which contributed towards the work for which he was awarded the Nobel Prize in Physics for 1920. Its discovery was particularly significant because of both its utility in high precision equipment, and its theoretical interest. Due to the nature of the alloying components it was soon determined to be related to a suppression of the anharmonic increase in atomic vibrations due to magnetic correlations; however the specific nature of these interactions is yet to be determined. A number of different proposed models are discussed by Rancourt [35]. As has been shown in the case of nickel, iron, and cobalt, magnetic correlations persist above the Curie temperature and into the liquid state [44]. Investigating these magnetic correlations in the liquid, as well as the local atomic configuration, could influence crystalline theoretical models.

Alongside work relating directly to the application of aerodynamic levitation is the consideration of the possibility of combining in situ aerodynamic levitation and anomalous neutron scattering (ANS) [48]. The feasibility of this combination is initially tested purely with the application of ANS to the structure of a liquid semiconductor, indium selenide.

The indium selenide system exhibits some interesting conductivity behaviour in the liquid state, including a wide compositional range of good semiconductivity, and an anomalous local maximum at In_2Se_3 [33]. As the short range order has a large influence on the semiconductivity in the liquid [15], it is important to determine the atomic structure in order to establish the origin of this behaviour. As InSe has recently been suggested as a possible candidate for phase change memory, its liquid structure is also of interest from a technological perspective [24].

As well as investigating liquid InSe, determining the experimental challenges of an ANS experiment on a sample in a container will give some indication of the viability of an ANS aerodynamic levitation experiment.

This thesis will be organised in the following manner:

As a significant theme of this thesis is the application of aerodynamic levitation to glass science, Chap. 2 contains a discussion of the characteristics of glass and a summary of theoretical attempts to describe the glassy state. Other pertinent theoretical descriptions, such as the theory of liquid semiconductors and the Invar problem, will be found in the relevant experimental chapters.

Chapter 3 will explain the four main experimental techniques used in this thesis; neutron diffraction, X-ray absorption spectroscopy (XAS), and computer simulations for structural studies, and aerodynamic levitation and laser heating for sample preparation.

The key experimental results of this thesis, which were discussed above, are presented in Chaps. 4, 5, 6, and 7. Initially each topic will be put in the wider context of the literature, before the relevant experimental and analytical methods are covered. Finally each result will be discussed. Chapter 8 then summarises the essential conclusions and discusses possible future implications.

References

1. Aasland S, McMillan PF (1994) Density-driven liquid–liquid phase separation in the system Al2O3-Y2O3. Nature 369:633–636
2. Akishige Y, Fukano K, Shigematsu H (2003) New ferroelectric BaTi2O5. Japanese J Appl Phys 42(8A):L946–L948 (Part 2)
3. Anderson PW (1995) Through the glass lightly. Science 267(5204):1615–1616
4. Angell CA (1995) The old problems of glass and the glass transition, and the many new twists. Proc Natl Acad Sci USA 92(15):6675–6682
5. Arai K, Namikawa H, Kumata K, Honda T, Ishii Y, Handa T (1986) Aluminum or phosphorus co-doping effects on the fluorescence and structural properties of neodymium-doped silica glass. J Appl Phys 59(10):3430–3436
6. Arai Y, Aoyama T, Yoda S (2004) Spherical sapphire single-crystal synthesis by aerodynamic levitation with high growth rate. Rev Sci Instrum 75(7):2262
7. Barnes AC, Skinner LB, Salmon PS, Bytchkov A, Pozdnyakova I, Farmer TO, Fischer HE (2009) Liquid–liquid phase transition in supercooled yttria-alumina. Phys Rev Lett 103(22):225702
8. Cohen I, Ha A, Zhao X, Lee M, Fischer T, Strouse MJ, Kivelson D (1996) A low-temperature amorphous phase in a fragile glass-forming substance. J Phys Chem 100(20):8518–8526
9. Coté B, Massiot D, Taulelle F, Coutures J-P (1992) 27Al NMR spectroscopy of aluminosilicate melts and glasses. Chem Geol 96(3–4):367–370
10. Cuello G, Fernández-Perea R, Bermejo FJ, Román-Ross G, Campo J (2007) Structure of Fe–Ni and Fe–Ni–S molten alloys by neutron diffraction. J Non-Cryst Solids 353(32–40):2987–2992
11. Debenedetti PG (2003) Supercooled and glassy water. J Phys: Condens Matter 15(45):R1669–R1726
12. Demirjian BG, Dosseh G, Chauty A, Ferrer M-L, Morineau D, Lawrence C, Takeda K, Kivelson D, Brown S (2001) Metastable solid phase at the crystalline-amorphous border: the glacial phase of triphenyl phosphite. J Phys Chem B 105(11):2107–2116
13. Dianoux A-J, Lander G (eds) (2003) Neutron data booklet, 2nd edn. Old City Publishing, Philadelphia
14. Enderby JE, North DM, Egelstaff PA (1966) The partial structure factors of liquid Cu-Sn. Phil Mag 14(131):961–970
15. Enderby JE, Barnes AC (1990) Liquid semiconductors. Rep Prog Phys 53(2):85–179
16. Fischer HE, Barnes AC, Salmon PS (2006) Neutron and X-ray diffraction studies of liquids and glasses. Rep Prog Phys 69(1):233–299
17. Greaves GN, Sen S (2007) Inorganic glasses, glass-forming liquids and amorphizing solids. Adv Phys 56(1):1–166
18. Greaves GN, Wilding MC, Fearn S, Langstaff D, Kargl F, Cox S, Van QVu, Majérus O, Benmore CJ, Weber R, Martin CM, Hennet L (2008) Detection of first-order liquid/liquid phase transitions in yttrium oxide-aluminum oxide melts. Science 322:566–570
19. Guillaume CE (1919–1920) The anomaly of the nickel-steels. Proc Phys Soc Lond 32:374–404
20. Guthrie M, Urquidi J, Tulk C, Benmore C, Klug D, Neuefeind J (2003) Direct structural measurements of relaxation processes during transformations in amorphous ice. Physical Review B 68(18):1–5
21. Hansen J-P, McDonald IR (1990) Theory of simple liquids, 2nd edn. Academic Press Limited, London
22. Katayama Y, Mizutani T, Utsumi W, Shimomura O, Yamakata M, Funakoshi K (2000) A first-order liquid–liquid phase transition in phosphorus. Nature 403(6766):170–173
23. Landron C, Hennet L, Jenkins T, Greaves G, Coutures J-P, Soper A (2001) Liquid alumina: detailed atomic coordination determined from neutron diffraction data using empirical potential structure refinement. Phys Rev Lett 86(21):4839–4842
24. Lee H, Kim YK, Kim D, Kang D-H (2005) Switching behavior of indium selenide-based phae-change memory cell. IEEE Trans Magn 41(2):1034–1036

25. March NH, Tosi MP (2002) Introduction to liquid state physics. World Scientific, Singapore
26. Masuno A, Inoue H, Yu J, Arai Y, Atsubo F (2008) Thermal stability and optical properties of Er3+ doped BaTi2O5 glasses. Adv Mater Res 39–40:243–246
27. Mishima O, Calvert LD, Whalley E (1984) Melting ice I at 77 K and 10 kbar: a new method of making amorphous solids. Nature 310:393–395
28. Monaco G, Falconi G, Crichton W, Mezouar M (2003) Nature of the first-order phase transition in fluid phosphorus at high temperature and pressure. Phys Rev Lett 90(25):255701
29. Nagashio K, Kuribayashi K (2002) Spherical yttrium aluminum garnet embedded in a glass matrix. J Am Ceram Soc 85(9):2353–2358
30. Nasikas NK, Sen S, Papatheodorou GN (2011) Structural nature of polymorphism in Y_2O_3-Al_2O_3 glasses. Chem Mater 23:2860–2868
31. Neilson GW, Adya AK (1996) Neutron diffraction studies on liquids. Annu Rep Prog Chem Sect C: Phys Chem 93:101–145
32. Nordine PC, Atkins RM (1982) Aerodynamic levitation of laser-heated solids in gas jets. Rev Sci Instrum 53(9):1456
33. Okada T, Ohno S (1993) Electrical properties of liquid In-Se alloys. J Non-Cryst Solids 156–158:748–751
34. Petri I, Salmon PS, Fischer H (2000) Defects in a disordered world: the structure of glassy $GeSe_2$. Phys Rev Lett 84(11):2413–2416
35. Rancourt DG (2002) Invar behavior in Fe-Ni alloys is predominantly a local moment effect arising from the magnetic exchange interactions between high moments. Phase Transitions 75(1–2):201–209
36. Rapoport E (1967) Model for melting-curve maxima at high pressure. J Chem Phys 46(8):2891–2895
37. Rapoport E (1967) Melting curve of $NaClO_3$. J Chem Phys 46(9):3279–3281
38. Rulison AJ, Watkins JL, Zambrano B (1997) Electrostatic containerless processing system. Rev Sci Instrum 68(7):2856
39. Skinner LB, Barnes AC, Crichton W (2006) Novel behaviour and structure of new glasses of the type Ba-Al-O and Ba-Al-Ti-O produced by aerodynamic levitation and laser heating. J Phys: Condens Matter 18(32):L407–L414
40. Skinner LB, Barnes AC, Salmon PS, Crichton WA (2008) Phase separation, crystallization and polyamorphism in the Y2O3-Al2O3 system. J Phys: Condens Matter 20:205103
41. Squires GL (1978) Introduction to the theory of thermal neutron scattering. Cambridge University Press, Cambridge
42. Weber JKR, Hampton DS, Merkley DR, Rey CA, Zatarski MM, Nordine PC (1994) Aero-acoustic levitation: a method for containerless liquid-phase processing at high temperatures. Rev Sci Instrum 65(2):456
43. Weber JKR, Abadie JG, Key TS, Hiera K, Nordine PC, Waynant RW, Ilev IK (2002) Synthesis and optical properties of rare-earth–aluminum oxide glasses. J Am Ceram Soc 85(5):1309–1311
44. Weber M, Knoll W, Steeb S (1978) Magnetic small-angle scattering from molten elements iron, colbalt and nickel. J Appl Crystallogr 11:638–641
45. Weber R, Hampton S, Nordine PC, Key T, Scheunemann R (2005) Er3+ fluorescence in rare-earth aluminate glass. J Appl Phys 98(4):043521
46. Wilding MC, Wilson M, McMillan PF (2005) X-ray and neutron diffraction studies and MD simulation of atomic configurations in polyamorphic Y2O3-Al2O3 systems. Phil Trans Ser A Math Phys Eng Sci 363(1827):589–607
47. Wouch G, Frost RT, Lord AE (1977) Preliminary observations of crystallization of levitated tungsten. J Cryst Growth 37(2):181–183
48. Wright AC, Cole JM, Newport RJ, Fisher CE, Clarke SJ, Sinclair RN, Fischer HE, Cuello GJ (2007) The neutron diffraction anomalous dispersion technique and its application to vitreous Sm2O3·4P2O5. Nucl Instrum Methods Phys Res Sect A 571(3):622–635
49. Yao K, Zhang LY, Yao X, Zhu WG (1997) Preparation and properties of barium titanate glass-ceramics sintered from sol-gel-derived powders. J Mater Sci 32(14):3659–3665

Chapter 2
Theory of Supercooled Liquids and Glasses

In this chapter I will attempt to describe the phenomenological understanding of supercooling and the glass transition. A glass is generally defined [25, 44] as an amorphous solid that has experienced a glass transition. This obviously raises the question of what constitutes a glass transition, which is a subject that requires significantly more discourse.

2.1 Kinetics of the Glass Transition

In order to correctly understand the changes which occur during the glass transition, we must first establish certain properties in both the equilibrium liquid and metastable supercooled liquid.

In the equilibrium liquid (above the melting point T_m) structural rearrangements occur due to diffusion, D, which, in the case of a low Reynolds number fluid with spherical particles, is related to the viscosity, η, by the Stokes-Einstein equation:

$$D = \frac{k_B T}{6\pi \eta R} \tag{2.1}$$

where k_B is Bolztmann's constant, T is the temperature, and R is the particle radius. In most cases [44] the viscosity is well described by the Arrhenius equation:

$$\eta(T) = \eta_0 \exp\left(\frac{E}{k_B T}\right) \tag{2.2}$$

where η_0 is a material dependent constant, and E is an activation energy. Along with the average structural relaxation time, many other relaxation times, such as dielectric relaxation, also have an Arrhenius temperature dependence. For clarity, a relaxation time is defined as the time taken for a system to return to equilibrium after a perturbation. At temperatures above the melting point the time dependence

T. Farmer, *Structural Studies of Liquids and Glasses Using Aerodynamic Levitation*,
Springer Theses, DOI: 10.1007/978-3-319-06575-5_2,

© Springer International Publishing Switzerland 2015

of relaxation processes is a simple Debye exponential, which is equivalent to $\beta = 1$ in the Kohlrausch-Williams-Watts (KWW) [73] equation:

$$\phi(t) = \exp\left(-(t/\tau)^{\beta}\right) \tag{2.3}$$

where τ a characteristic relaxation time, and $\varphi(t)$ is the relaxation function [15], which is related to the measured quantity $\sigma(t)$ by:

$$\phi(t) = [\sigma(t) - \sigma(\infty)]/[\sigma(0) - \sigma(\infty)]. \tag{2.4}$$

It can be shown that due to the unique properties of exponential behaviour, the relaxation has no dependence on the previous behaviour of the system; the relaxation time is determined only by the temperature the system is at, not by previous temperatures [32].

By contrast, once the liquid is in the metastable supercooled state the relaxation becomes non-exponential [5], and is often well described by the complete KWW Eq. (2.3) with $0 < \beta < 1$. The value of β depends not only upon the material in question but also on the specific relaxation process. Along with the change in time dependent behaviour, the temperature dependence of the average relaxation time can also deviate strongly from the Arrhenius behaviour shown above. This deviation is also seen in the viscosity, and is often well described, albeit over only a few orders of magnitude [5], by the Vogel-Tammann-Fulcher (VTF) [21, 64, 70] equation:

$$\eta(T) = \eta_0 \exp\left(\frac{E}{k_B(T - T_0)}\right) \tag{2.5}$$

where T_0 is the temperature at which the viscosity (or relaxation) is predicted to become infinite. The extent to which the viscosity deviates from Arrhenius behaviour can vary quite significantly, as shown in Fig. 2.1. This leads to a classification of glass forming liquids depending on the extent to which their viscosity becomes non-Arrhenius in the supercooled region; strong liquids, where SiO_2 is the archetype, show little deviation from Arrhenius behaviour, whereas fragile liquids, such as o-terphenyl (OTP) show a marked deviation and are better described by the VTF equation. However, despite the improved description of fragile liquids by the VTF equation relative to Arrhenius behaviour, it is still only sufficient over a few decades of viscosity [5]. Hodge [32] determined that there is a correlation between the fragility of a glass-forming liquid and the extent to which its relaxation function is non-exponential.

From this it can be seen that as the temperature of a supercooled liquid is being reduced, the time it takes for structural relaxations to occur increases very rapidly. If the cooling rate is kept constant, then at a certain temperature there will no longer be sufficient time for the liquid to return to equilibrium before the temperature is reduced. At this temperature the liquid undergoes a kinetic transition to the glassy phase and the structure is effectively fixed on an experimental timescale. This is often defined [25, 44] as ~100 s, which corresponds to a viscosity of ~10^{12}Pa s. It is also apparent from this definition of the glass transition that it is dependent on the rate at which the liquid is cooled; a slower cooling rate will

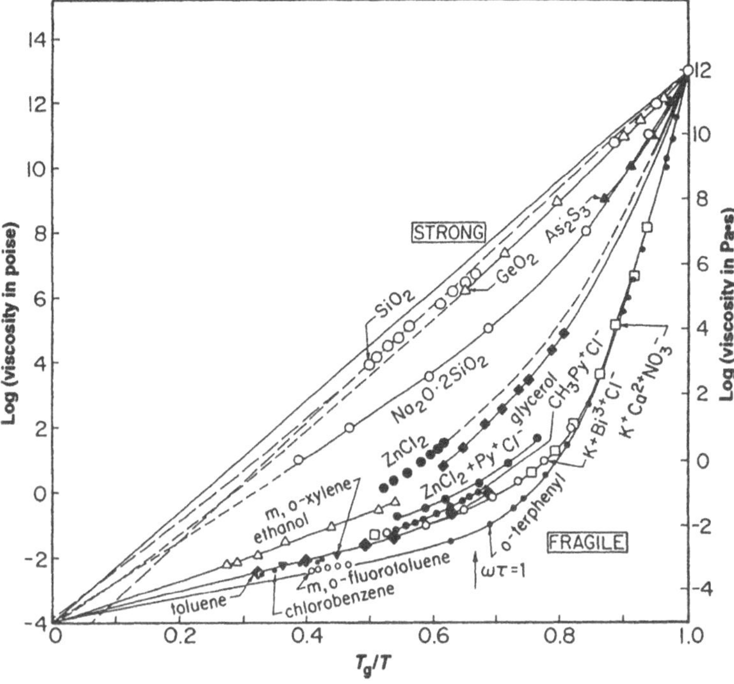

Fig. 2.1 A graph showing the wide range of viscosity behaviours of glass forming liquids as the temperature, T, approaches the glass transition temperature, T_g. (Reproduced in part from Angell [4], reprinted with permission from AAAS. Originally from Angell [3], © 1991, with permission from Elsevier.)

allow the liquid to retain its metastable equilibrium to a lower temperature than a faster cooling rate. Therefore the glass transition temperature is actually a range, determined by the cooling rate. However because of the rapid increase in viscosity with a reduction in temperature, the glass transition temperature is only weakly dependent on the cooling rate; Ediger et al. [18] suggest a temperature difference of 3–5 K with an order of magnitude change in the cooling rate. From both of these points it is apparent that the glass transition temperature is actually an arbitrary temperature range based on both the quench rate and the time over which we measure structural relaxations. The latter of these points is highlighted in Fig. 2.2 which shows the temperature dependence of the frequency dependent specific heat capacity in glycerol [8].

The change to super-Arrhenius relaxations in the supercooled regime only affects some relaxation processes, generally termed α relaxations [62]. In fragile glasses other relaxation processes (β processes) have been observed to retain their Arrhenius nature, and so at a certain temperature there is a splitting of relaxation processes [15]. This can be seen in the case of dielectric relaxation in Fig. 2.3. The specific mechanism behind this splitting and the resulting β relaxations is not well established [22, 56].

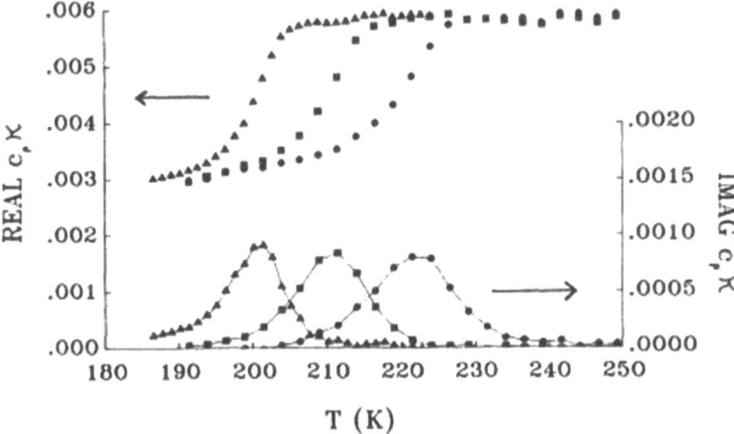

Fig. 2.2 The real and imaginary product of the frequency dependent specific heat capacity and the thermal conductivity for glycerol, as determined by Birge and Nagel [8], © 1985 by the American Physical Society. The three different frequencies are 0.62 Hz (*triangles*), 34 Hz (*squares*), and 1,100 Hz (*circles*). The glass transition temperature for glycerol, as determined by differential scanning calorimetry, is 180 K. A sharp decrease in specific heat capacity is indicative of a glass transition, as discussed in Sect. 2.2

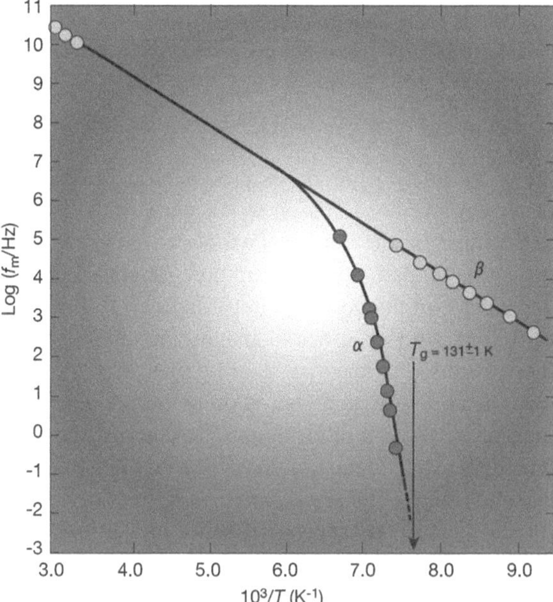

Fig. 2.3 The splitting of the peak dielectric relaxation frequency of a chlorobenzene/cis-decalin mixture into two separate relaxations. Figure reproduced from Debenedetti and Stillinger [15], Reprinted by permission from Macmillan Publishers Ltd: © 2001)

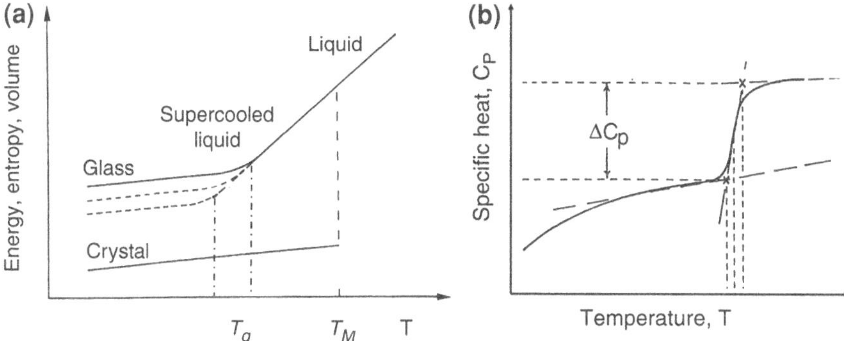

Fig. 2.4 A schematic showing the change in (**a**) the first and (**b**) the second derivatives of the free energy at the glass transition. The *dash-dotted lines* in (**a**) highlight that T_g occurs over a range of temperatures dependent on the quench rate. It is important to note that while the specific heat capacity in (**b**) changes sharply, it is not discontinuous (Adapted from Greaves and Sen [25], Reprinted with permission from Taylor and Francis Ltd. http://www.informaworld.com)

2.2 Thermodynamics of the Glass Transition

While so far I have only discussed the kinetics of the glass transition, it is important to also consider the thermodynamics. At the glass transition temperature range there is a significant change in the thermodynamic properties. This is clearly seen in both the first and second derivatives of the free energy, such as the entropy and the specific heat capacity, as shown schematically in Fig. 2.4. Despite the fact that over the range there is a large change in the specific heat capacity, it is observable that there is no discontinuity [4]. This eliminates the possibility that the glass transition is either a first order or second order phase transition as understood by the Ehrenfest classification [19]. Although Fig. 2.4 does not show a discontinuity in the entropy indicative of a first order phase transition, it does highlight an interesting paradox that was first established by Kauzmann [37]. As the specific heat capacity of the liquid is higher than that of the solid states, its entropy decreases more rapidly with decreasing temperature. Therefore if the metastable liquid state can be maintained without either crystallization or glass formation intervening, there will be a temperature, T_K, at which the entropy of the liquid decreases below that of the corresponding crystal. Although this is not forbidden in itself, it creates the possibility that the entropy could drop to zero at $T > 0$ K, thus violating the third law of thermodynamics. There are a few possible resolutions to Kauzmman's paradox, one of which involves the formation of an 'ideal glass', which would of course have a solid-like heat capacity, at T_K. This concept hints at the possibility of a true thermodynamic phase transition, and it has been suggested [4] that the kinetic glass transition that is experimentally observed could be masking this. In the case of fragile liquids, where there is often a large change in the specific heat capacity upon crystallization, T_K is often not very far below T_g. Despite this it has not been experimentally possible to access T_K without the intervention of either vitrification or crystallization.

2.3 Theoretical Descriptions of the Viscous Slowdown

For completeness some mention should be made of a few of the theoretical descriptions of the increase in relaxation times as the glass transition is approached.

The approach of Cohen and Turnbull [12, 68] was to state the necessary condition for a molecular rearrangement to occur in terms of a 'free volume'. The origin of this was related to the observation of a marked decrease in fluidity (or increase in viscosity) with applied pressure. By considering the diffusion as requiring motion into a free volume of critical size (assuming no hop back occurs), they reduced the problem of the glass transition to the determination of the temperature dependent free volume distribution. In this description the glass transition occurs when the energy required to redistribute the free volume becomes large.

Adam and Gibbs [2] chose instead to describe a configuration in terms of the fraction of cooperatively rearranging regions; these are regions of a specific size within which a configurational rearrangement can take place effectively independently of the macroscopic system. Consideration of the temperature dependent transition probability allows the determination of a minimum critical size of these regions, below which the transition probability is zero. As this critical size increases with decreasing temperature, there exists a temperature at which it diverges, which equates to the glass transition. It is important to note that both the free volume theory [12] and the configuration entropy theory [2] have derived equivalent equations to the VTF equation.

The final approach that will be discussed is that of mode coupling theory (MCT) [7, 23, 40]. It was pointed out [7] that in the liquid state the time dependent density correlation function tends to zero, as after a sufficiently long time diffusion has caused the density at any point to be independent of that after time t. However in the glassy state if we assume there is no diffusion then the density correlations do not decay. The key concept of MCT is an attempt to describe the approach to this glassy state by means of a non-linear density correlation function; this acts in such a way that the density correlation function has a memory. This non-linearity couples the size of the density fluctuations with their ability (through the longitudinal viscosity) to decay; therefore as the density fluctuations increase, the viscosity increases, which in turn reduces the decay of the density fluctuations. It is important to note that the original versions of MCT [7] resulted in a divergence which was not observed; this has since been corrected by modifications (see [15]).

2.4 Avoided Crystallization

It is also instructive to consider glass formation from the perspective of avoided crystallization. In order for a crystal structure to form, either heterogeneous or homogeneous nucleation must occur. Heterogeneous nucleation takes place at sites that are preferential to nucleation, such as seed crystals or solid surfaces. By contrast, homogenous nucleation is the formation of crystalline nuclei purely from statistical atomic rearrangements. Below a certain volume these nuclei are unstable and will tend back to the general liquid structure. However once a certain critical volume is achieved it becomes energetically favourable for crystal to grow [69]. This occurs when the decrease in free energy due to increasing the crystal volume overcomes the increase in free energy required to form the crystal-liquid interface, which is dependent on the surface area as opposed to the volume. Due to the temperature dependence of the free energy difference required for nucleation, the critical size reduces as the liquid is more deeply supercooled. However a competing factor occurs in the form of the decreasing viscosity, which reduces the ability for atoms to be added to a nucleus. Once a crystal nucleus does reach critical volume, crystallization progresses rapidly [67].

2.5 Glass Structure

While the mathematical formalism for describing the atomic structure will only be introduced in the following chapter, I will first briefly mention the common structural motifs of glasses.

Although glasses are amorphous they are found to have a significant degree of short range order and even intermediate range order [60] (greater than nearest neighbour ordering). This structure was initially described by Zachariasen [75] through the use of a continuous random network (CRN), as shown in Fig. 2.5a. Zachariasen [75] suggested that the simplest glasses, such as the archetypal glass SiO_2, were due to relatively small modifications of the corresponding crystal structure, maintaining the basic polyhedra but requiring a wide distribution of inter-polyhedral bond angles; this restriction on the polyhedra leads to the intermediate range order described by the CRN. Expanding on this description, [24] suggested an adaptation to allow for the inclusion of network modifying cations, such as alkalis or alkaline earths (as opposed to Si which is a network former). This modified random network (Fig. 2.5b) posits the possibility of modifier channels throughout the glass network, which have been suggested by EXAFS studies [24]. These modifier channels are surrounded by non-bridging oxygen atoms, which are oxygen atoms that are not bonded to two network former atoms.

The description of oxygen atoms as bridging (BO) or non-bridging (NBO) is an important concept in the oxide glasses, as it allows some insight into the depolymerization of the glass forming network. This in turn relates to the glass transition temperature and the fragility [4]. In the archetypal case of silicate glasses, Q^n is commonly used [25] to represent n BO atoms around each Si atom, with $0 \leq n \leq 4$ due to the tetrahedral coordination of Si. While the determination of the Q-species distribution is not directly obtainable from diffraction measurements (although it is from RMC from diffraction data), it is achievable through nuclear magnetic resonance (NMR) studies. Due to this capability and the ability for coordination number distribution determination in a number of cases [17], NMR is often used in conjunction with diffraction measurements [13].

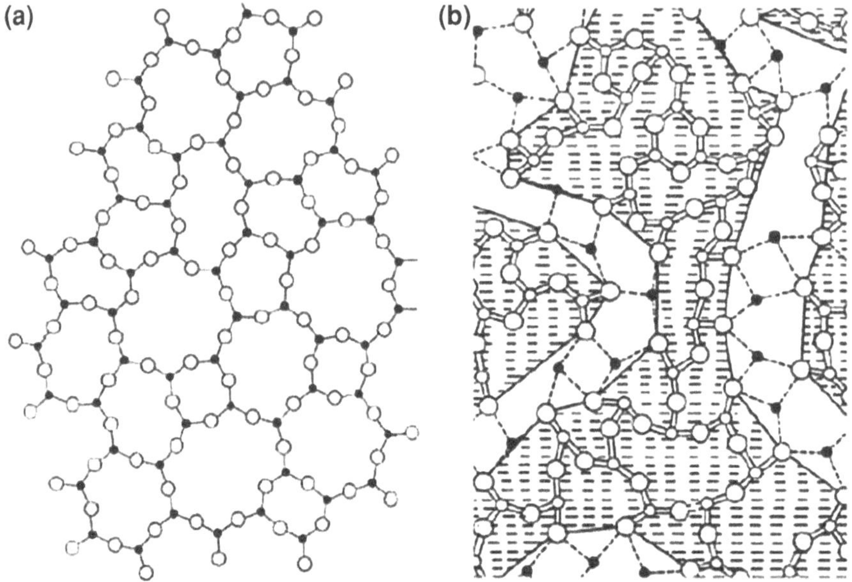

Fig. 2.5 **a** The continuous random network as described by Zachariasen (Reprinted with permission from [75], ©1932 American Chemical Society.) where silicon atoms (*closed circles*) are connected by bridging oxygen atoms (*open circles*). **b** The modified random network of Greaves (Reprinted from [24], © 1985, with permission from Elsevier.) where modifier channels occur due to the addition of alkalis or alkaline earths (*closed circles*) to the Si (*small open circles*) and O (*large open circles*) network

Although it is possible to have a range of different polyhedra [25], the most commonly occurring is the tetrahedra (Fig. 2.6). Tetrahedra are the main structural unit of a number of glasses, including many oxide, chalcogenide, and halide glasses. They can connect in three different ways (Fig. 2.6): corner sharing such as SiO_2 [25]; edge sharing such as $SiSe_2$ [44]; and face sharing such as Ge-Se-AgI [9]; or in combinations of these such as $GeSe_2$ [44].

2.6 Polyamorphism and First Order Liquid–Liquid Phase Transitions

When discussing liquid–liquid phase transitions it is important to establish how exactly they are being defined. In this particular case I am defining a liquid–liquid transition as being iso-compositional and first order in nature (i.e. there is a discontinuous change in the first derivatives of the free energy, such as the specific heat capacity at constant pressure or the density). As the two phases must have the same Gibbs free energy at the phase transition they must be able to coexist. This means that we arrive at the picture of a liquid phase of one density nucleating inside a liquid phase with a different density, as opposed to the usual idea

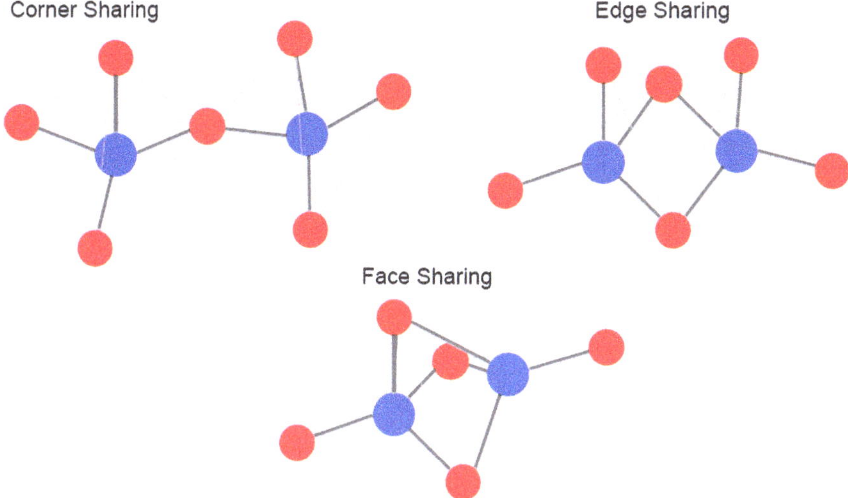

Fig. 2.6 The common structural motif in glasses, the tetrahedra, shown with the three different connecting configurations

of a continuous change in the density as one of the thermodynamic variables are altered. Given the situation of one liquid nucleating within another, glassy polyamorphism occurs if the sample is then quenched at such a rate that not only do both phases preferentially form a glass instead of a crystal, but also that this occurs before the phase transition causes a complete change from one liquid phase to the other. This would therefore manifest itself as small inclusions of one amorphous phase inside a matrix of the second amorphous phase.

The concept of a first order liquid–liquid phase transition is based upon the explanation of melting curve maxima of [58, 59] in 1967. Such melting curve maxima (Fig. 2.7) have been observed in a wide variety of different types of liquids, including Cs [38], P [35], Te [39], Ba, $NaClO_3$ [59], and H_2O [50], showing that it is a behaviour not related to a specific type of bonding.

The theory goes as follows: in the stable liquid state there are two distinct structural species of the liquid, a high density liquid (HDL) and a low density liquid (LDL). Due to fluctuations each component will vary between the two species, with only the time averaged distribution of species staying constant for a specific temperature and pressure. Upon an increase in pressure the proportion of the more dense species increases, which leads to an increase in the density of the liquid above that of the corresponding crystal. This causes the melting curve to have a negative relationship with pressure, as can be seen from the Clausius-Clapeyron relationship:

$$\frac{dT_m}{dP} = \frac{\Delta V_m}{\Delta S_m} \tag{2.6}$$

where T_m is the melting temperature, P is the pressure, ΔV_m is the change in volume upon melting, and ΔS_m is the change in entropy. As the change in entropy is always positive upon melting, an increase in the density upon melting means dT_m

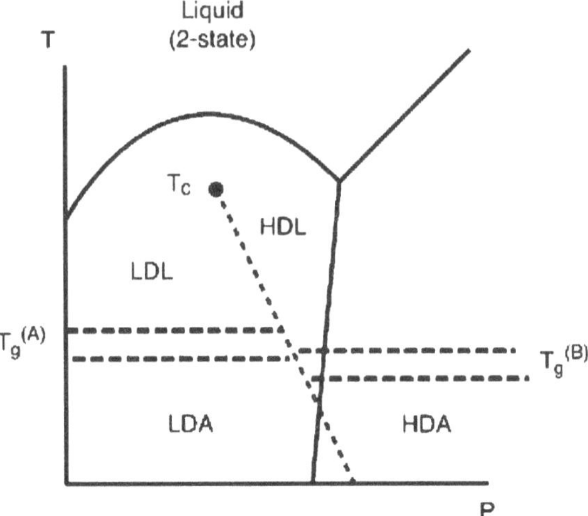

Fig. 2.7 A temperature-pressure diagram of a liquid with a melting curve maximum (Reproduced in part from McMillan [46] with permission of The Royal Society of Chemistry). The LDL vitrifies in the region denoted by $T_g^{(A)}$ to a low density amorphous phase (*LDA*), whereas the HDL vitrifies in a separate region, $T_g^{(B)}$, to a distinct high density amorphous phase (*HDA*)

/ $dP < 0$. In instances such as the phase diagram of water, where the melting curve does not display an initial maximum, it is predicted that this occurs in the metastable tensile-strained regime where the equilibrium state is the vapour.

Adapting Guggenheim's theory of binary solutions [27], Rapoport [58] determined that there is a critical value leading to a separation into two liquids when:

$$\frac{w}{k_B T} > 2 \tag{2.7}$$

where w is the energy cost for replacing one species with another, and k_B is Boltzmann's constant. This led to the suggestion of a first order transition between a higher density liquid and a lower density liquid, whilst maintaining the same composition. The enthalpy change that occurs in a liquid–liquid transition is expected to be lower than that during a solid-liquid transition, as the degree of atomic rearrangement is expected to be lower changing between two disordered structures than in nucleating an ordered crystal. From this and the inequality above (2.7) it is probable that, at a specific pressure, the liquid–liquid transition temperature is below the melting temperature. This would mean that in general the liquid–liquid phase transition line would be expected to occur in the supercooled regime, ending with a critical point below T_m.

Firstly I will highlight in detail the particular cases of liquid–liquid phase transitions in water and yttria aluminates, before briefly mentioning a few other examples.

2.6.1 Water

Until this point I have only discussed glass formation from the perspective of quenching a liquid, and while this is both the oldest and most commonly used method, we must also consider pressure induced amorphisation (PIA). PIA for was initially reported by Mishima et al. [50] by cooling crystalline ice Ih to 77 K before pressurising to ~1 GPa. This produced a significant, apparently discontinuous, change in volume over a small pressure range, which was then shown to be amorphous using X-ray diffraction. It is important to note that [50] suggested the X-ray diffraction data showed the new amorphous phase was not the structure that results from rapidly cooling liquid water [45], and that the crystal to amorphous transition was not reversible.

Expanding on this discovery the same authors [51] investigated the effects of heating this new amorphous ice (which will now be termed high density amorphous or HDA). They apparently determined that heating HDA to ~117 K at atmospheric pressure caused a sharp transition to a low density amorphous structure (~22 % lower density), although they only suggest that it "strongly resembles a first-order transition".

While there are three main theoretical explanations [14] for the anomalous behaviour of H_2O, I will only be focusing on the possibility of a liquid–liquid phase transition ending in a critical point. This was initially suggested by Poole et al. [57] based on both the experimental studies discussed above [51], and their own MD simulations. These simulations used the ST2 [63] potential for water, although similar effects have been seen using the TIP4P [34] potential. As pointed out by Mishima and Stanley [47], and Franzese et al. [20], the occurrence of a first order liquid–liquid phase transition is a natural consequence of a double welled potential (Fig. 2.8a), such as ST2. They also highlight the fact that, due to the complexity of the real interaction potentials, water is notoriously difficult to simulate accurately [47]. This is evident when considering that some simulations with the ST2 potential have predicted three liquid–liquid critical points [10].

The position in *PT* space of the proposed [57] liquid–liquid transition is below the homogenous nucleation temperature for supercooled water (Fig. 2.8b). This means that it is very difficult to experimentally investigate the region of the possible liquid–liquid phase transition. Due to this, Mishima and Stanley [48] adopted a method utilising the melting curve of metastable ice IV, which happens to pass through the anticipated liquid–liquid phase transition line. At the predicted crossover they detected a change in the melting behaviour indicating that the ice IV was melting to a different liquid, before crystallizing to the stable crystalline ice phase. Despite this hint of a possible liquid–liquid transition, the results could be influenced by a few different factors; the experimental setup used very small quantities (droplets of ~1 μm diameter) of water as an emulsion rather than pure H_2O, which raises questions about the effects of both the confinement and the surface interactions; the measured crossover is very close to the homogenous nucleation temperature, which may result in a similar effect; and the separation of data points raises the possibility of whether there is a sharp, but not continuous transition [14].

There is a significant degree of contention as to whether the transition between LDA and HDA is discontinuous [14, 28, 29, 66]. Although initial studies [51, 52] of the compression with applied pressure suggest the transition was sharp, as does a Raman study by Mishima and Suzuki [49], there have been a significant number of X-ray [29] and neutron diffraction studies [66], and MD simulations [28] that have implied that the structural changes actually occur continuously between

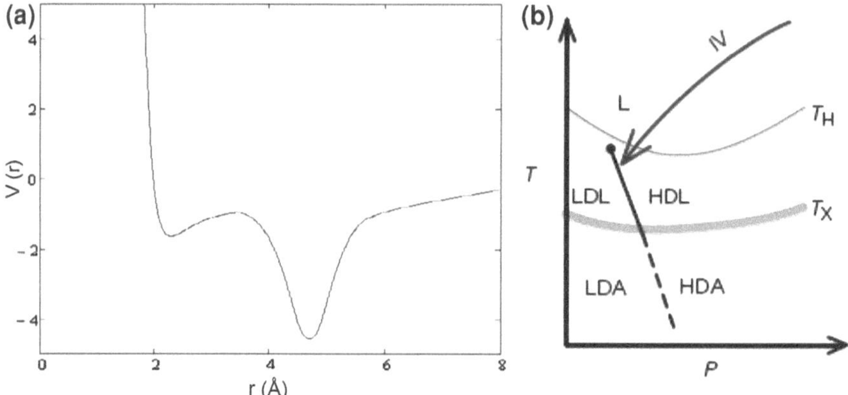

Fig. 2.8 **a** A schematic of a double welled potential. At high temperatures both wells can be sampled. As the temperature reduces the deeper, low density well becomes the preferentially filled, resulting in a LDL. As the pressure is increased the system becomes situated in the less deep well, corresponding to a higher density liquid. **b** The proposed position in *TP* space of the liquid–liquid phase transition in water, reproduced from Mishima and Stanley (Reprinted by permission from Macmillan Publishers Ltd: [48], © 1998). T_H is the homogenous nucleation temperature line for supercooled water, while T_X is the temperature at which amorphous ice spontaneously crystallizes upon heating

an infinite number of intermediary states. A recent MD study by Limmer and Chandler [41] questions the appearance of a liquid–liquid transition in MD simulations of one component liquids, by suggesting that what is being observed is in fact a liquid-solid transition.

The possibility of the existence of a second liquid–liquid phase transition in water was discussed following the discovery [42] of an even higher density amorphous state, named very high density amorphous (VHDA). However, unlike the LDA-HDA transition, this has since been widely accepted to be a continuous change in the structure [74].

When considering the range of conflicting results and the experimental challenges, it is easy to understand why the existence of a first order liquid–liquid phase transition in water has not been conclusively established.

2.6.2 Al_2O_3-Y_2O_3 System

The initial reports of a liquid–liquid phase transition in the Al_2O_3-Y_2O_3 system were due to Aasland and Mcmillan's [1] (AM) observation of polyamorphism in a concentration range of 24–32 mol% Y_2O_3. They produced samples using a combination of sol-gel synthesis followed by Ir wire heating. The volume fraction of the reported glassy inclusions increased with both decreasing Y_2O_3 content and quench rate; it is important to note that crystallization occurred in samples of 20 and 22 mol% Y_2O_3. In order to determine the amorphous nature of the inclusions AM (1994) used micro-infrared and Raman spectroscopy. However, as pointed out by Nagashio and Kuribayashi [55] (NK), due to the transparency of the samples it is difficult to ensure that spectra are taken from inclusions located on the surface, rather than some thickness of bulk glass separating an inclusion from the surface. Using aero-acoustic levitation, NK (2002) produced AY samples across the same composition which showed the same morphology of an inclusion containing matrix as AM (1994). Using microfocus XRD they demonstrated that the inclusions present on the surface of their sample had sharp peaks consistent with crystalline YAG.

Further evidence to suggest that the observation of amorphous inclusions in the Al_2O_3-Y_2O_3 system is in fact that onset of crystallization, comes from the microfocus EXAFS of Barnes et al. [6]. Although similar in principal to both the micro-infrared and Raman spectroscopy of AM (1994) and the microfocus XRD of NK (2002), this method had the significant advantage that inclusions on the surface showed strong optical fluorescence, definitively establishing that the spectrum came from an inclusion. Further characterisation of these inclusions using cross-polarised microscopy also determined that they were crystalline of nature [61].

Along with the possible glass polyamorphism in Al_2O_3-Y_2O_3 previously discussed, Greaves et al. [26] observed an apparent liquid–liquid phase transition in situ using aerodynamic levitation combined with small and wide angle X-ray scattering (SAXS/WAXS). In addition to this they also suggested that a repeating 0.25 Hz oscillation in the pyrometry was due to a ‘polyamorphic rotor’. These observations will be discussed further in Chap. 5, when compared with in situ SANS experiments and pyrometric studies.

Similarly to water, MD simulations [71] have also suggested the possibility of a liquid–liquid phase transition in the yttria-aluminates.

2.6.3 Other Examples

There are a number of other systems which are possible candidates for demonstrating liquid–liquid phase transitions.

Triphenyl phosphate (TPP) is an organic compound of $P(OC_6H_5)_3$ which Cohen et al. [11, 30] first suggested exhibited polyamorphism in 1996. TTP liquid can be supercooled, although if it is quenched slowly it crystallizes at ~245 K. However if it is fast quenched below ~225 K it is stable against crystallization for extended periods of time. If it is kept in this supercooled state between 213 and 225 K for a matter of hours (during which it becomes turbid, opaque, and then finally clear again) it appears to form a ‘glacial phase’ [11] of apparently different density, although the X-ray diffraction is very similar [30]. The ‘glacial phase’ is so called because it is apparently glassy [11] but is distinct from the TPP glass formed by fast quenching below 176 K. As with both water and the yttria-aluminate system there are a number of different views [31, 33, 65] on what phenomenon is manifesting itself. A number of these come under the general theme of an apparently disordered phase that contains nanocrystallites [16].

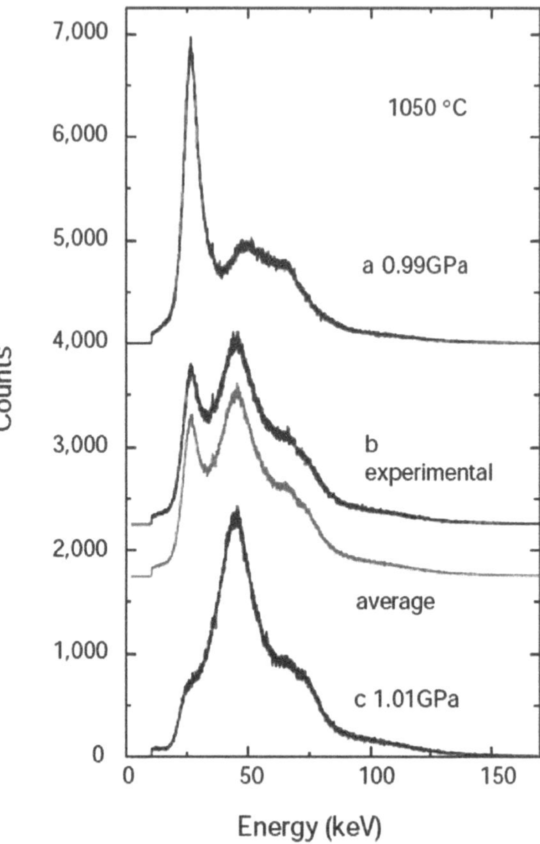

Fig. 2.9 The X-ray data of Katayama et al. (Reprinted by permission from Macmillan Publishers Ltd: [35], © 2000) which initially suggested a first order liquid–liquid phase transition in phosphorus. The *grey line* represents a weighted average of the two distinct species observed at high and low pressure, and accurately recreates the experimental data taken between these two pressures. This strongly suggested coexistence which is an essential requirement for a first order transition

In 2000 Katayama et al. [35] conducted an in situ X-ray diffraction experiment on liquid phosphorus which strongly suggested a first order liquid–liquid phase transition between a molecular liquid and a polymeric liquid. This was particularly notable because not only did it occur in the stable liquid, as opposed to the more common metastable supercooled liquid, it also demonstrated coexistence of the two phases. This was clearly visible from a weighted sum of the diffraction patterns from either side of the transition compared to the diffraction pattern at the transition (Fig. 2.9). The coexistence was also apparent from in situ X-ray radiography [36] and had been suggested by ab initio MD simulations [54]. However a further, more extensive, in situ X-ray diffraction investigation by Monaco et al. [53] established that the transition was in fact between a polymeric liquid and a molecular fluid; the latter being the fluid form associated with the metastable white phosphorus crystal, which takes a molecular tetrahedral structure. Therefore while undoubtedly a first order liquid-fluid phase transition, phosphorus did not provide the first evidence for a first order liquid–liquid phase transition.

It has also been suggested [25] that polyamorphism could occur, or has been already detected, in some network glasses, such as SiO_2-GeO_2 [43]. However, again due to experimental challenges [72] relating to the proposed location in the phase diagram of the polyamorphism, it has been difficult to conclusively establish.

To summarise, there has been significant amount of research into the possibility of a first order isocompositional liquid–liquid phase transition. So far there has been no overwhelming evidence in support of this hypothetical explanation of negative melting curves in any system. The results of [26] strongly suggest the existence of a liquid–liquid phase transition in the yttria aluminates; it is therefore essential that this discovery is independently established, as will be discussed in Chap. 5.

References

1. Aasland S, McMillan PF (1994) Density-driven liquid–liquid phase separation in the system Al_2O_3-Y_2O_3. Nature 369:633–636
2. Adam G, Gibbs JH (1965) On the temperature dependence of cooperative relaxation properties in glass-forming liquids. J Chem Phys 43(1):139–146
3. Angell CA (1991) Relaxation in liquids, polymers and plastic crystals—strong/fragile patterns and problems. J Non-Cryst Solids 131–133:13–31
4. Angell CA (1995) Formation of glasses from liquids and biopolymers. Science 267(5206):1924–1935
5. Angell CA, Ngai KL, Mckenna GB, McMillan PF, Martin SW (2000) Relaxation in glass forming liquids and amorphous solids. J Appl Phys 88(6):3113–3157
6. Barnes AC, Skinner LB, Salmon PS, Bytchkov A, Pozdnyakova I, Farmer TO, Fischer HE (2009) Liquid–liquid phase transition in supercooled yttria-alumina. Phys Rev Lett 103(22):225702
7. Bengtzelius U, Götze W, Sjölander A (1984) Dynamics of supercooled liquids and the glass transition. J Phys C: Solid State Phys 17(33):5915–5934
8. Birge NO, Nagel SR (1985) Specific-heat spectroscopy of the glass transition. Phys Rev Lett 54(25):2674–2677
9. Boev V, Mitkova M, Lefterova E, Wagner T, Kasap S, Vlček M (2000) Glass formation in the Ge ± Se ± AgI ternary. J Non-Cryst Solids 266:867–871
10. Brovchenko I, Geiger A, Oleinikova A (2003) Multiple liquid–liquid transitions in supercooled water. J Chem Phys 118(21):9473–9476
11. Cohen I, Ha A, Zhao X, Lee M, Fischer T, Strouse MJ, Kivelson D (1996) A Low-temperature amorphous phase in a fragile glass-forming substance. J Phys Chem 100(20):8518–8526
12. Cohen MH, Turnbull D (1959) Molecular transport in liquids and glasses. J Chem Phys 31(5):1164–1169
13. Cole JM, van Eck ERH, Mountjoy G, Newport RJ, Brennan T, Saunders GA (1999) A neutron diffraction and ^{27}Al MQMAS NMR study of rare-earth phosphate glasses, $(R_2O_3)_x(P_2O_5)_{1-x}$, x = 0.187-0.263, R = Ce, Nd, Tb containing Al impurities. J Phys Condens Matter 11(47):9165–9178
14. Debenedetti PG (2003) Supercooled and glassy water. J Phys Condens Matter 15(45):R1669–R1726
15. Debenedetti PG, Stillinger FH (2001) Supercooled liquids and the glass transition. Nature 410(6825):259–267
16. Demirjian BG, Dosseh G, Chauty A, Ferrer M-L, Morineau D, Lawrence C, Takeda K, Kivelson D, Brown S (2001) Metastable solid phase at the crystalline-amorphous border: the glacial phase of triphenyl phosphite. J Phys Chem B 105(11):2107–2116
17. Eckert H (1992) Structural characterization of noncrystalline solids and glasses using solid state NMR. Prog Nucl Magn Reson Spectrosc 24(3):159–293

18. Ediger MD, Angell CA, Nagel SR (1996) Supercooled liquids and glasses. J Phys Chem 100(31):13200–13212
19. Finn CBP (1993) Thermal physics, 2nd edn. Chapman and Hall, London
20. Franzese G, Malescio G, Skibinsky A, Buldyrev SV, Stanley HE (2001) Generic mechanism for generating a liquid–liquid phase transition. Nature 409(6821):692–695
21. Fulcher GS (1925) Analysis of recent measurements of the viscosity of glasses. J Am Ceram Soc 8:339
22. Goldstein M (2011) The past, present, and future of the Johari-Goldstein relaxation. J Non-Cryst Solids 357(2):249–250
23. Götze W, Sjögren L (1992) Relaxation processes in supercooled liquids. Rep Prog Phys 55(3):241–376
24. Greaves GN (1985) EXAFS and the structure of glass. J Non-Cryst Solids 71:203–217
25. Greaves GN, Sen S (2007) Inorganic glasses, glass-forming liquids and amorphizing solids. Adv Phys 56(1):1–166
26. Greaves GN, Wilding MC, Fearn S, Langstaff D, Kargl F, Cox S, Van QVu, Majérus O, Benmore CJ, Weber R, Martin CM, Hennet L (2008) Detection of first-order liquid/liquid phase transitions in yttrium oxide-aluminum oxide melts. Science 322:566–570
27. Guggenheim EA (1935) The Statistical Mechanics of Regular Solutions. Proc R Soc A Math Phys Eng Sci 148(864):304–312
28. Guillot B, Guissani Y (2003) Polyamorphism in low temperature water: a simulation study. J Chem Phys 119(22):11740
29. Guthrie M, Urquidi J, Tulk C, Benmore C, Klug D, Neuefeind J (2003) Direct structural measurements of relaxation processes during transformations in amorphous ice. Phys Rev B 68(18):1–5
30. Ha A, Cohen I, Zhao X, Lee M, Kivelson D (1996) Supercooled liquids and polyamorphism. J Phys Chem 100(1):1–4
31. Hédoux A, Guinet Y, Descamps M, Hernandez O, Derollez P, Dianoux AJ, Foulon M, Lefèbvre J (2002) A description of the frustration responsible for a polyamorphism situation in triphenyl phosphite. J Non-Cryst Solids 307:637–643
32. Hodge IM (1994) Enthalpy relaxation and recovery in amorphous materials. J Non-Cryst Solids 169(3):211–266
33. Johari GP, Ferrari C (1997) Calorimetric and dielectric investigations of the phase transformations and glass transition of triphenyl phosphite. J Phys Chem 5647(49):10191–10197
34. Jorgensen WL, Chandrasekhar J, Madura JD, Impey RW, Klein ML (1983) Comparison of simple potential functions for simulating liquid water. J Chem Phys 79(2):926
35. Katayama Y, Mizutani T, Utsumi W, Shimomura O, Yamakata M, Funakoshi K (2000) A first-order liquid–liquid phase transition in phosphorus. Nature 403(6766):170–173
36. Katayama Y, Inamura Y, Mitzutani T, Yamakata M, Utsumi W, Shimomura O (2004) Macroscopic separation of dense fluid phase and liquid phase of phosphorus. Science 306(5697):848–851
37. Kauzmann W (1948) The nature of the glassy state and the behavior of liquids at low temperatures. Chem Rev 43(2):219–256
38. Kennedy GC, Jayaraman A, Newton RC (1959) Fusion curve and polymorphic transitions of cesium at high pressures. Phys Rev 126(4):1363–1366
39. Klement W Jr, Cohen LH, Kennedy GC (1966) Melting and freezing of selenium and tellurium at high pressures. J Phys Chem Solids 27(1):171–177
40. Leutheusser E (1984) Dynamical model of the liquid-glass transition. Phys Rev A 29(5):2765
41. Limmer DT, Chandler D (2011) The putative liquid–liquid transition is a liquid-solid transition in atomistic models of water. J Chem Phys 135(13):134503
42. Loerting T, Salzmann C, Kohl I, Mayer E, Hallbrucker A (2001) A second distinct structural "state" of high-density amorphous ice at 77 K and 1 bar. Phys Chem Chem Phys 3(24):5355–5357
43. Majérus O, Cormier L, Itié J-P, Galoisy L, Neuville DR, Calas G (2004) Pressure-induced Ge coordination change and polyamorphism in SiO_2–GeO_2 glasses. J Non-Cryst Solids 345–346:34–38
44. March NH, Tosi MP (2002) Introduction to liquid state physics. World Scientific, Singapore

45. Mayer E, Brüggeller P (1982) Vitrification of pure liquid water by high pressure jet freezing. Nature 298(5876):715–718
46. McMillan PF (2004) Polyamorphic transformations in liquids and glasses. J Mater Chem 14(10):1506–1512
47. Mishima O, Stanley HE (1998) The relationship between liquid, supercooled and glassy water. Nature 396(6907):329–335
48. Mishima O, Stanley HE (1998) Decompression-induced melting of ice IV and the liquid–liquid transition in water. Nature 392(6672):164–168
49. Mishima O, Suzuki Y (2002) Propagation of the polyamorphic transition of ice and the liquid–liquid critical point. Nature 419(6907):599–603
50. Mishima O, Calvert LD, Whalley E (1984) "Melting ice" I at 77 K and 10 kbar: a new method of making amorphous solids. Nature 310:393–395
51. Mishima O, Calvert LD, Whalley E (1985) An apparently first-order transition between two amorphous phases of ice induced by pressure. Nature 314(6006):76
52. Mishima O, Takemura K, Aoki K (1991) Visual observations of the amorphous–amorphous transition in H_2O under pressure. Science 254(5030):406–408
53. Monaco G, Falconi G, Crichton W, Mezouar M (2003) Nature of the first-order phase transition in fluid phosphorus at high temperature and pressure. Phys Rev Lett 90(25):255701
54. Morishita T (2001) Liquid–liquid phase transitions of phosphorus via constant-pressure first-principles molecular dynamics simulations. Phys Rev Lett 87(10):105701
55. Nagashio K, Kuribayashi K (2002) Spherical yttrium aluminum garnet embedded in a glass matrix. J Am Ceram Soc 85(9):2353–2358
56. Paluch M, Roland C, Pawlus S, Zioło J, Ngai K (2003) Does the Arrhenius temperature dependence of the Johari-Goldstein relaxation persist above T_g? Phys Rev Lett 91(11):115701
57. Poole PH, Sciortino F, Essmann U, Stanley HE (1992) Phase behaviour of metastable water. Nature 360(6402):324–328
58. Rapoport E (1967) Model for melting-curve maxima at high pressure. J Chem Phys 46(8):2891–2895
59. Rapoport E (1967) Melting curve of $NaClO_3$. J Chem Phys 46(9):3279–3281
60. Salmon PS (2007) The structure of tetrahedral network glass forming systems at intermediate and extended length scales. J Phys: Condens Matter 19(45):455208
61. Skinner LB, Barnes AC, Salmon PS, Crichton WA (2008) Phase separation, crystallization and polyamorphism in the Y_2O_3-Al_2O_3 system. J Phys: Condens Matter 20:205103
62. Stillinger FH (1995) A topographic view of supercooled liquids and glass formation. Science 267(5206):1935–1939
63. Stillinger FH, Rahman A (1974) Improved simulation of liquid water by molecular dynamics. J Chem Phys 60(4):1545–1557
64. Tammann G, Hesse W (1926) Die abhängigkeit der viskosität von der temperatur bei unterkühlten flüssigkeiten. Z Anorg Allg Chem 156:245–257
65. Tanaka H, Kurita R, Mataki H (2004) Liquid–liquid transition in the molecular liquid triphenyl phosphite. Phys Rev Lett 92(2):025701
66. Tse J, Klug D, Guthrie M, Tulk C, Benmore C, Urquidi J (2005) Investigation of the intermediate- and high-density forms of amorphous ice by molecular dynamics calculations and diffraction experiments. Phys Rev B 71(21):214107
67. Turnbull D (1969) Under what conditions can a glass be formed? Contemp Phys 10(5):473–488
68. Turnbull D, Cohen MH (1961) Free-volume model of the amorphous phase: glass transition. J Chem Phys 34(1):120–125
69. Turnbull D, Fisher JC (1949) Rate of nucleation in condensed systems. J Chem Phys 17(1):71–73
70. Vogel H (1921) Das temperatur-abhängigkeitsgesetz der viskosität von flüssigkeiten. Phys Zeit 22:645–646
71. Wilding MC, Wilson M, McMillan PF (2005) X-ray and neutron diffraction studies and MD simulation of atomic configurations in polyamorphic Y_2O_3-Al_2O_3 systems. Philosophical transactions. Series A, Mathematical, physical, and engineering sciences 363(1827):589–607

72. Wilding M, Guthrie M, Bull CL, Tucker MG, McMillan PF (2008) Feasibility of in situ neutron diffraction studies of non-crystalline silicates up to pressures of 25 GPa. J Phys Condens Matter 20(24):244122
73. Williams G, Watts C (1970) Non-symmetrical dielectric relaxation behaviour arising from a simple empirical decay function. Trans Faraday Soc 66(565P):80–85
74. Winkel K, Elsaesser M, Mayer E, Loerting T (2008) Water polyamorphism: reversibility and (dis)continuity. J Chem Phys 128(4):044510
75. Zachariasen WH (1932) The atomic arrangement in glass. J Am Chem Soc 54:3841–3851

Chapter 3
Experimental Techniques

3.1 Aerodynamic Levitation

3.1.1 Overview

This technique involves the use of a stream of gas directed through a conical nozzle in order to support a small sample (~1.5–5 mm diameter, 10–100 mg). Typically, high purity (99.9995 %) argon is used as a levitation gas in order to reduce the possibility of contamination; however other levitation gases can also be used, for instance an argon/oxygen mix if the oxygen content of the sample is likely to reduce with Ar as the sole levitation gas. Once a sample is levitating, it is then heated with a 10.6 μm 240 W continuous wave CO_2 laser in order to induce melting (Fig. 3.1). Contactless temperature measurement is achieved using either a single wavelength or multi wavelength pyrometer. In the case of a single wavelength pyrometer the relationship between the true temperature, T, and the apparent temperature, T_a, due to the emissivity, ε, is [35]:

$$\frac{1}{T} - \frac{1}{T_a} = \frac{\lambda}{C_2} \ln\left(\varepsilon_\lambda\right) \tag{3.1}$$

where λ is the wavelength of the pyrometer and C_2 is Planck's second radiation constant. In the case of most oxides the emissivity in the infrared is sufficiently high [41] that there is only a small deviation (~2 %) in temperature from that of a blackbody. However in the case of metals, where the reflectivity is high, the emissivity could be as small as 0.2, causing a large deviation (~600 K at a temperature of 2,300 K) in the actual sample temperature from that measured by the pyrometer. Dual wavelength pyrometers avoid the requirement of knowledge of the emissivity, although they rely on the assumption that the emissivity is wavelength independent. They are usually also limited to slower response times than single wavelength pyrometers. It is important to note that due to the observable release of latent heat even for small amounts of crystallization, pyrometry is the first indication of successful vitrification.

T. Farmer, *Structural Studies of Liquids and Glasses Using Aerodynamic Levitation*,
Springer Theses, DOI: 10.1007/978-3-319-06575-5_3,

© Springer International Publishing Switzerland 2015

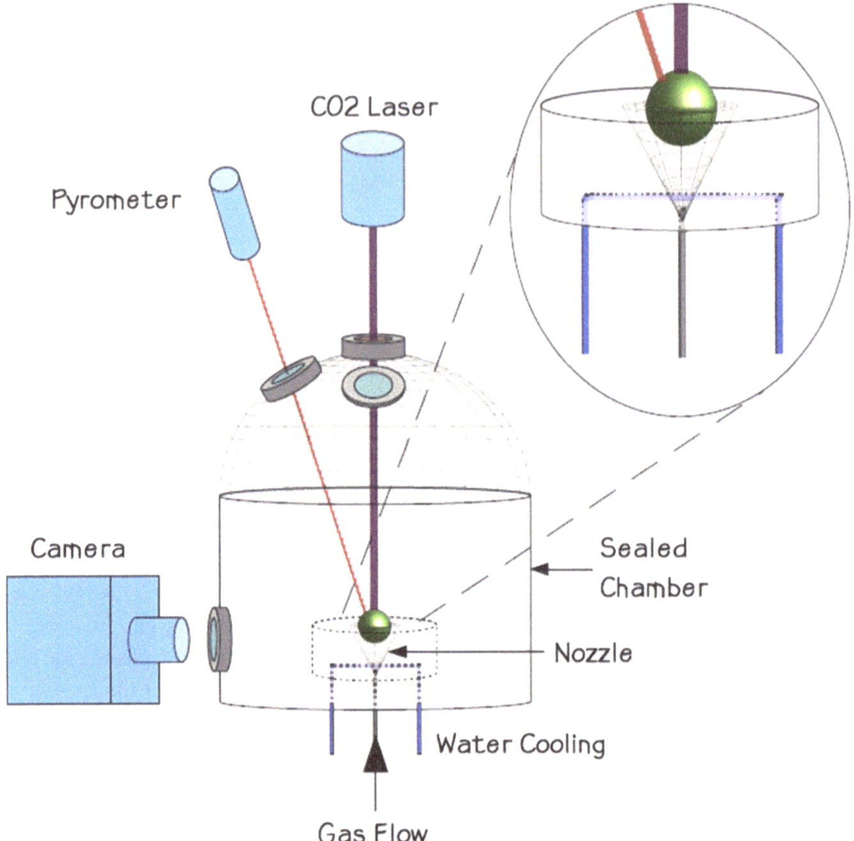

Fig. 3.1 A schematic of the aerodynamic levitator setup used for this work. The laser and pyrometer are not incident on the same part of the sample

For the work in this thesis a 1.15 μm single wavelength pyrometer with a response time of 30 μs and a spot size of ~0.25 mm was used.

As shown in Fig. 3.1 the pyrometry is measured at the top hemisphere of the sample. As the laser is incident upon the top of the sample the pyrometry only indicates the maximum temperature of the sample. Depending on the size of the sample, its thermal conductivity, viscosity, and transmission, the temperature gradient from the top of the sample is expected to be as large as 100 K [43]. It should be mentioned that laser heating has been applied to the top and bottom of samples simultaneously [30].

Levitation methods have several advantages over conventional furnace heating. The possibility of crucible contamination, which is especially significant with high temperature melting points, is eliminated. Along with this is the removal of external sites of heterogeneous nucleation, which suppresses the possibility of crystallization when producing glass. Another advantage with respect to investigating glass formation is the high degree of control over the quench rate. As the laser can either be instantly cut-off (quench rate ~400 K/s although this has a significant sample size dependence) or ramped down at any rate, it is possible to investigate how varying the quench rate affects the glass morphology.

A limitation of aerodynamic levitation is an incompatibility with very low density samples (<2 g/cm^3) [14]. Along with this are the drawbacks of laser heating, which is not suitable for melting some samples. This can occur for a number of reasons; due to a combination of very high T_m and high reflectivity; when the thermal conductivity is sufficiently low that the sample is ablated (such as for sulphide glasses); or when the sample does not absorb at 10.6 μm (such as for ZnSe glass).

For samples where the mass loss during heating is minimal, it is possible to achieve stable levitation for periods in excess of 24 h. This allows the possibility of in situ measurements of various different liquid properties, such as electrical conductivity [42], density [52], and structural measurements [16, 27, 32]. In the case of diffraction this requires an aerodynamic levitator setup on either a neutron or X-ray diffraction beamline, such as on D4c [32] at the ILL or D2AM [27] at the ESRF.

The aerodynamic levitator used for the work in this thesis was purchased from Containerless Research, Inc., Evanston, IL, USA.

3.1.2 Sample Production

Levitation samples were produced using two different methods involving laser fusing of powders on a copper hearth.

The first of these, dubbed the two ball method by Skinner [41], is suitable for producing a small number of samples of very high compositional accuracy. This method consists of using laser heating on a copper hearth to fuse a set of solid spheres consisting of a single powder constituent (e.g. Al_2O_3). A second set of spheres of the second powder constituent (e.g. Y_2O_3) are then fused, and the mass of each ball is compared with the desired masses based upon the masses of the first set and the intended compositions. If a match is found within a small mass range (usually less than 0.5 %), the two balls are then fused and levitated for a minimum of 60 s. Due to the relatively low viscosity [24] and small size of the samples this allows sufficient mixing of the constituent components to occur. As the mass lost during the final fusing is significantly lower than fusing from a powder (usually less than 0.1 %), this usually ensures a compositional accuracy of better than 1 %. Obviously this method is just as applicable for samples consisting of three or more constituent components, although in this case would be more accurately named the multi-ball method.

As the two ball method is very time intensive it is not suitable for producing large quantities (>30) of samples of the same composition. In this case it is more suitable to initially measure the correct mass of each constituent for ~1.5 g total mass. The combined powders are then pressed into a pellet and baked into a solid sample in a box furnace. Following this the solid pellet is ground up using a pestle and mortar for approximately 30 min. After this process has been completed it is assumed that the powder is a homogeneous mixture of the constituents and solid spheres of this powder are then laser fused on the copper hearth before levitation. Samples produced by this method have been checked for compositional accuracy using energy dispersive X-ray spectroscopy (EDX) and found to be compositionally accurate within the limits of EDX (<1 %). It should be highlighted that this technique is of limited use for compositional analysis of oxygen. A further check when producing glasses is that the known glass forming range limits the compositional variation.

3.2 Neutron Scattering from Amorphous Materials

3.2.1 Overview

While a full discussion of the theory of neutron scattering is beyond the scope of this work, it is important to summarise the key results. Initially I will discuss the common formalism for neutron diffraction applied to amorphous structures. Following this I will consider two methods for determining information at the partial structure factor level; isotopic substitution, and the use of absorption resonances for anomalous neutron scattering (ANS). Due to its application in Chap. 5 a brief description of small angle neutron scattering (SANS) will be given. Similarly, because of the paramagnetic nature of the samples in Chaps. 4 and 6, neutron scattering from unpaired electrons is also discussed. Finally the correction methods that were applied to the raw data are described. A detailed review of neutron and X-ray scattering from amorphous materials is given by Fischer et al. [23].

3.2.2 Nuclear Scattering

The geometry of a neutron diffraction experiment is shown in Fig. 3.2, where a neutron of incident wavevector $\mathbf{k}_i$ is scattered into a state with wavevector $\mathbf{k}_f$. In the case of nuclear scattering, the scattering centre is the atomic nucleus. The essential information that is determined in a neutron (or X-ray) diffraction measurement is the differential scattering cross-section:

$$\frac{d\sigma}{d\Omega}(\mathbf{q}) = \frac{\text{number of neutrons scattered per second into } d\Omega}{I_0\, d\Omega} \tag{3.2}$$

where I_0 is the incident flux of neutrons, and $d\Omega$ is the solid angle into which scattering occurs (Fig. 3.2).

Initially we will make the static approximation [48] and assume that the energy exchanged in scattering is small compared with the incident energy. This is equivalent to assuming that the time for a neutron to travel a distance of the

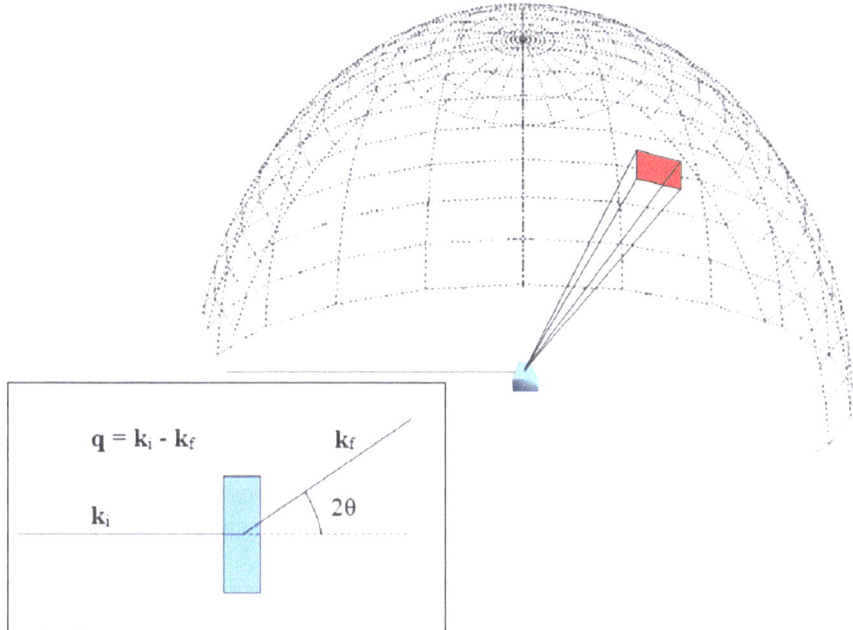

Fig. 3.2 The geometry of a neutron scattering experiment, with the scattering *solid angle*, $d\Omega$, shown in *red*. The *inset* shows a top down view of the sample with the incident and final wavevectors, $\mathbf{k}_i$ and $\mathbf{k}_f$, and the scattering angle 2θ

order of an atomic separation is small compared with the time for atomic motion [23]. The validity of this approximation for neutron scattering from liquids is insufficient, which leads to corrections discussed later in Sect. 3.2.7.3.

Neutron scattering from a monatomic system can have variations in the bound scattering length, b, from atom to atom. These differences can occur due to a system being comprised of several spinless isotopes, a single isotope with non-zero spin, or a combination of the two. It is important to note that b is complex and can be negative, which is equivalent to an attractive Fermi pseudopotential [48] and therefore a π phase change in the wave. If we assume that there is no correlation between atomic position and b then the cross section can be split into two contributions:

$$\begin{aligned} \left(\frac{d\sigma}{d\Omega}(\mathbf{q})\right) &= \left(\frac{d\sigma}{d\Omega}(\mathbf{q})\right)_{distinct} + \left(\frac{d\sigma}{d\Omega}(\mathbf{q})\right)_{self} \\ &= \overline{b}^2 \sum_{\substack{i,j \\ i \neq j}}^{N} \exp\left(i\mathbf{q} \cdot \mathbf{r}_{ij}\right) + N\overline{b^2} \end{aligned} \tag{3.3}$$

where $\mathbf{r}_{ij}$ is the separation of the two atoms:

$$\mathbf{r}_{ij} = \mathbf{r}_i - \mathbf{r}_j \tag{3.4}$$

From this it is readily apparent that only the distinct term contains structural information, while the self term is an isotropic contribution to the differential scattering cross section that is purely dependent on the average squared scattering length. Often the differential scattering cross section is instead split into coherent and incoherent parts:

$$\begin{aligned}\left(\frac{d\sigma}{d\Omega}(\mathbf{q})\right) &= \left(\frac{d\sigma}{d\Omega}(\mathbf{q})\right)_{coh} + \left(\frac{d\sigma}{d\Omega}(\mathbf{q})\right)_{incoh} \\ &= \bar{b}^2 \sum_{i,j}^{N} \exp\left(i\mathbf{q}\cdot\mathbf{r}_{ij}\right) + N\left(\overline{b^2} - \bar{b}^2\right)\end{aligned} \tag{3.5}$$

The reason for this is that the static structure factor is defined as:

$$S(\mathbf{q}) = \sum_{i,j}^{N} \exp\left(i\mathbf{q}\cdot\mathbf{r}_{ij}\right) \tag{3.6}$$

The static structure factor can be Fourier transformed into real space to give the pair correlation function, $g(\mathbf{r})$, which is proportional to the probability of finding an atom at a position $\mathbf{r}$ relative to a reference atom at the origin. Due to the isotropic nature of liquids and glasses it is only necessary to consider the distance from the reference atom; hereafter the vector dependence becomes a magnitude dependence. Therefore the pair correlation function, $g(r)$, is expressed in terms of $S(q)$:

$$g(r) - 1 = \frac{1}{\rho_0 2\pi^2 r} \int_0^\infty q\left[S(q) - 1\right] \sin(qr) dq \tag{3.7}$$

with the inverse relationship being:

$$S(q) - 1 = \frac{4\pi\rho_0}{q} \int_0^\infty r\left[g(r) - 1\right] \sin(qr) dr \tag{3.8}$$

The pair correlation function naturally introduces the concept of a coordination number; the average number of atoms that are situated in the spherical shell between two distances r_1 and r_2 centred on a reference atom:

$$n = 4\pi\rho_0 \int_{r_1}^{r_2} g(r) r^2 dr \tag{3.9}$$

Obviously in the case of amorphous structures, which are by definition disordered, there is no necessity for the coordination number to be integer; this illustrates that the different members of same atomic species can occur in multiple coordinations simultaneously. A coordination number is only strictly defined in the case that $g(r_2) = 0$, although it is common to quote coordination numbers to the first minimum even if this remains finite.

Two important consistency checks can be determined from the relationship between $g(r)$ and $S(q)$. From (3.7) we can determine the intuitive result that

$g(r \to \infty) = 1$, as the probability of finding any atom at any large distance r becomes equal. Similarly from (3.8) we see that $S(q \to \infty) = 1$. Further to this is the other limit of the structure factor [23, 48]:

$$S(0) = \rho_0 \kappa_T k_B T \tag{3.10}$$

where κ_T is the isothermal compressibility, k_B is Boltzmann's constant and T is the absolute temperature.

The above derivation can be generalised to a polyatomic system with one additional consideration. While in a monatomic system the distribution of scattering lengths can be considered random, in polyatomic systems the atoms generally occupy preferred sites. So although for any particular chemical species α, the distribution of α isotopes over the α sites is random, there is an overall correlation between scattering length and atomic position. Due to this the relation of the total interference contribution, $F(q)$, to the total differential scattering cross section (i.e. excluding the isotropic scattering) is a weighted combination of partial structure factors:

$$\begin{aligned} F(q) &= \frac{1}{N}\frac{d\sigma}{d\Omega}(q) - \sum_{\alpha}^{n} c_\alpha \overline{b_\alpha^2} \\ &= \sum_{\alpha,\beta}^{n} c_\alpha c_\beta \overline{b_\alpha b_\beta^*}\left[S_{\alpha\beta}(q) - 1\right] \end{aligned} \tag{3.11}$$

where c_α is the concentration of the chemical species α, and b^* is the complex conjugate if b. The partial structure factors are not unique and can be defined in different formalisms to suit the situation. In this thesis two formalisms will be used; that of Faber and Ziman [19] (FZ) which is defined in (3.11): and that of Bhatia and Thornton [6] (BT) which will be discussed below. The Fourier transform of the total structure factor leads to the total pair correlation function, $G(r)$, which is a weighted combination of the FZ partial pair correlation functions:

$$\begin{aligned} G(r) &= \frac{1}{2\pi^2 r \rho_0}\int_0^\infty qF(q)\sin(qr)dq \\ &= \sum_{\alpha,\beta} c_\alpha c_\beta \overline{b_\alpha b_\beta^*}\left(g_{\alpha\beta}(r) - 1\right) \end{aligned} \tag{3.12}$$

The partial pair correlation functions, $g_{\alpha\beta}(r)$, are just the probability of finding any atom of type α at a distance r from any central atom β. It should be noted that for the purposes of this thesis the concentration and scattering length factor in (3.12) will be referred to as the prefactor. The definition of a corresponding polyatomic coordination number, the average number of β atoms that are situated in the spherical shell between two distances r_1 and r_2 around an α atom, follows:

$$n_\alpha^\beta = 4\pi\rho_0 c_\beta \int_{r_1}^{r_2} g_{\alpha\beta}(r) r^2 dr \tag{3.13}$$

From (3.12) we also gain another useful consistency check, which is often a good method of ensuring the correct number density has been used:

$$G(0) = -\sum_{\alpha,\beta} c_\alpha c_\beta b_\alpha b_\beta \tag{3.14}$$

This is equivalent to the condition of the initial slope being equal to $4\pi\rho_0$ in the commonly used total correlation function [23].

As mentioned above, the other definition of the partial structure factors used in this thesis is the BT formalism [6, 39], which more directly relate to thermodynamic variables. In this formalism the partial structure factors, $S_{NN}(q)$, $S_{CC}(q)$, and $S_{NC}(q)$, are defined as follows:

$$\begin{aligned} F(q) &= |\langle b\rangle|^2 S_{NN}(q) + \left|\overline{b_1} - \overline{b_2}\right|^2 S_{CC}(q) \\ &\quad + \left[\langle b\rangle\left(\overline{b}_1^* - \overline{b}_2^*\right) + \langle b\rangle^*\left(\overline{b}_1 - \overline{b}_2\right)\right] S_{NC}(q) - \left(c_1\overline{b_1}^2 + c_2\overline{b_2}^2\right) \end{aligned} \tag{3.15}$$

where:

$$\langle b\rangle = c_1 b_1 + c_2 b_2 \tag{3.16}$$

In the case of the BT structure factors a little more explanation is required than for the FZ structure factors. $S_{NN}(q)$ is the number–number partial structure factor; from (3.15) we can see that this is purely related to the number density fluctuations, and that it is uniquely determined when scattering lengths of the two components are the same. $S_{CC}(q)$ is the concentration-concentration partial structure factor; its real space equivalent, $g_{CC}(r)$, gives the probability of finding atoms of the same species or of opposite species, at a distance r away from a central atom, given the atomic sites specified by $S_{NN}(q)$. Therefore if there is no preferential chemical ordering, $S_{CC}(q)$ has no q dependence. $S_{NC}(q)$ is the cross-correlation between number and concentration; the Fourier transform of the number-concentration partial structure factor, $g_{NC}(r)$, describes the probability at a distance r of a specific site being occupied by a specific chemical species.

3.2.3 Isotopic Substitution

From the discussion above it is easy to see that the greatest amount of structural information that can be determined from neutron scattering on an amorphous material is at the partial structure factor level. However it is also evident that, other than in the case of one or more null scatterers ($b_\alpha = 0$) [23], a single neutron diffraction measurement on a polyatomic system is only capable of determining $G(r)$. From (3.11) it is evident that in order to determine partial structure factors the scattering length of one or more component must be varied. Fortunately as discussed above, the scattering length varies with isotope, a clear example being hydrogen (b = −3.409 fm) and deuterium (b = 6.674 fm) [17]. The possibility of isotopic substitution was first explored by Enderby et al. [18], and relies on the reasonable assumption that there is no change in interatomic interactions with a change in isotope. For any polyatomic system, the number of different measurements that must be made for a complete determination of the partial structure factors is equal to the number of partial structure factors [23]. This fact is highlighted when (3.12) is expressed in matrix form, as is shown for a binary system:

$$\begin{bmatrix} F_1(q) \\ F_2(q) \\ F_3(q) \end{bmatrix} = \begin{bmatrix} c_\alpha^2 \bar{b}_{\alpha 1}^2 & c_\beta^2 \bar{b}_{\beta 1}^2 & 2c_\alpha c_\beta \bar{b}_{\alpha 1} \bar{b}_{\beta 1} \\ c_\alpha^2 \bar{b}_{\alpha 2}^2 & c_\beta^2 \bar{b}_{\beta 2}^2 & 2c_\alpha c_\beta \bar{b}_{\alpha 2} \bar{b}_{\beta 2} \\ c_\alpha^2 \bar{b}_{\alpha 3}^2 & c_\beta^2 \bar{b}_{\beta 3}^2 & 2c_\alpha c_\beta \bar{b}_{\alpha 3} \bar{b}_{\beta 3} \end{bmatrix} \begin{bmatrix} S_{\alpha\alpha}(q) - 1 \\ S_{\beta\beta}(q) - 1 \\ S_{\alpha\beta}(q) - 1 \end{bmatrix} \quad (3.17)$$

where $F_I(q)$ is the total structure factor from the measurement on α_I and β_I, and so on. In order to determine the partial structure factors from the totals, (3.17) is inverted. While it is obvious that a greater difference in each total structure factor will lead to a more precise determination of the partial structure factors, quantifying the required difference for a specific statistical uncertainty in $F(q)$ is more challenging. In this thesis a conditioning measure described by Westlake [51] is used. This equates to normalizing the matrix of (3.17) by dividing each row, I, by $(\sum_j a_{ij}^2)^{1/2}$ (where a_{ij} specifies an element of the matrix) and taking the determinant; the ideal case resulting in a determinant equal to 1. Fischer et al. [23] discuss this and a few other possible methods.

3.2.4 Nuclear Resonances

The neutron scattering length is a complex quantity, where the real part determines the scattering cross section and the imaginary part the absorption cross section:

$$b(E) = b_{real}(E) + ib_{imag}(E) \quad (3.18)$$

where the scattering cross section, σ_{scat}, and absorption cross section, σ_{abs}, are dependent on the real and imaginary parts respectively:

$$\sigma_{scat} = 4\pi b_{real}^2 \quad \sigma_{abs} = 2b_{imag}\lambda \quad (3.19)$$

For most elements the neutron scattering length at thermal neutron wavelengths is independent of the energy, so the energy dependence is often not explicitly stated in (3.18). In these cases the scattering cross section is constant and the absorption cross section has a linear dependence on the wavelength, as can be seen from (3.19). However some elements, such as In, Cd, and several lanthanides, have real or virtual absorption resonances at thermal neutron energies (~1 eV). These resonances cause variations in both b_{real} and b_{imag} which can be calculated from the Breit-Wigner [9, 15] equations:

$$b_{real}(E) = b_{real}(E_0) + 9.103 \times 10^3 \left(\frac{A+1}{A}\right)^2 \left[\sum_j g\Gamma_{nj}^0 \left(\frac{E - E_j}{4(E - E_j)^2 + \Gamma_j^2} - \frac{E_0 - E_j}{4(E_0 - E_j)^2 + \Gamma_j^2}\right.\right] \tag{3.20}$$

$$b_{imag}(E) = b_{imag}(E_0) + 4.551 \times 10^3 \left(\frac{A+1}{A}\right)^2 \left[\sum_j g\Gamma_{nj}^0 \left(\frac{1}{4(E - E_j)^2 + \Gamma_j^2} - \frac{1}{4(E_0 - E_j)^2 + \Gamma_j^2}\right)\right] \tag{3.21}$$

where A is the nuclear mass, g is the spin statistical factor, E_0 is the reference energy, E_j is the energy of the jth absorption resonance, Γ_j is total width of the jth resonance, and:

$$\Gamma_{nj}^0 = \Gamma_{nj}\sqrt{\frac{1\,\text{eV}}{E_j}} \tag{3.22}$$

is the reduced neutron width at 1 eV, where Γ_{nj} is the neutron width of the *jth* resonance.

In the case of b_{imag} these increases can be several orders of magnitude as the resonance energy is approached. This obviously causes a significant increase in the absorption cross section, which creates difficult experimental conditions.

For spallation sources, which simultaneously utilise a range of wavelengths, this energy dependence can hamper or prevent measurements on samples containing elements with resonances. However for reactor sources it is possible to take advantage of the variation in the scattering length in much the same way as anomalous X-ray scattering [23]. ANS, as first suggested by Krogh-Moe [31] and demonstrated by Wright et al. [53], employs at least two different neutron wavelengths in order to achieve the same effect as isotopic substitution; it is important to note again that in this case the imaginary contribution cannot be ignored. For a binary system three wavelengths are required in order to entirely separate the three correlation functions, and two wavelengths can provide a first order difference [23]. Similar to the case of isotopic substitution it is possible to use a null scatterer to determine a correlation function from a single wavelength.

3.2.5 Small Angle Neutron Scattering

When discussing SANS it is initially preferable to start by considering what constitutes the measured intensity [40]:

$$I(q) = I_0(\lambda)\eta(\lambda)T(\lambda)d\Omega\left(\frac{d\sigma}{d\Omega}(q)\right)_{SANS} \otimes R(q) \tag{3.23}$$

where again $I_0(\lambda)$ is the incident flux, $\eta(\lambda)$ is the detector efficiency, $T(\lambda)$ is the sample transmission, and $R(q)$ is the instrumental resolution function. In (3.23) the differential scattering cross section is modified for the SANS case [29]:

$$\frac{d\sigma}{d\Omega}(q) = N_p V_p^2 (\Delta\rho)^2 P(q) S(q) \tag{3.24}$$

where N_p is the number of scatterers, V_p is the volume of each scatterer, $P(q)$ is the form factor, and $\Delta\rho$ is the difference in scattering length densities (discussed further below). In this case the structure factor, $S(q)$, describes the interparticle structure, rather than interatomic. It is important to note that a scatterer need not be a well defined molecule; a $\Delta\rho$ originating from density fluctuations will cause an equivalent change in the differential scattering cross section.

The equivalency of the two forms of the differential scattering cross section is explained as follows. In the earlier discussion of neutron diffraction from amorphous substances the concept of scattering length was introduced. The scattering length is specific to neutron diffraction from nuclei, and is strictly part of a broader definition of functions which includes the atomic form factor in the case of X-rays, and the magnetic form factor for magnetic neutron diffraction. The general form of this function is commonly called the scattering length density distribution [40], and is dependent on the distribution of this interaction in real space, such as the atomic form factor being the Fourier transform of the electron density. For neutron diffraction from nuclei the interaction is of such short range that in q space it is effectively constant over required q ranges; thus the fact it can be described purely by a single number. In the case of an arbitrary particle in a matrix the real space interaction is dependent on the shape of the particle, the volume, and the difference in scattering between it and the matrix.

For the purposes of this thesis several simplifications can be made to (3.24). As the SANS I am concerned with is from the nucleation of an amorphous thermodynamic phase we can essentially consider the particles to be near spherical. Along with this we can assume that the collection of particles is sufficiently dilute that the correlations between them (and therefore static structure factor in (3.24)) can be ignored.

3.2.6 Magnetic Neutron Scattering

In the first section I considered purely the effects of a neutron scattering from a nucleus due to the strong nuclear force. However neutrons also posses a magnetic moment and so will also be scattered from unpaired electrons. This possibility was initially suggested by Bloch [8], who also pointed out that despite the vastly different magnitudes of the two forces responsible, the resultant scattering strength would be comparable. Unlike nuclear scattering however, because of the similar relative scales of the neutron wavelength and the radius of an electron cloud, and the fact that the electromagnetic force is long range, magnetic scattering is q dependent.

As in the nuclear case, only a summary of the relevant results from the theory of magnetic neutron scattering will be given; there are numerous texts that cover the topic in detail [3, 48]. In the context of this thesis, magnetic neutron scattering is only considered in terms of correcting for paramagnetic effects in liquid transition metals and rare earths.

As mentioned above, magnetic neutron scattering occurs due to the interaction of the neutron's magnetic moment with unpaired electrons of momentum **p** and spin **s** (if we discount the generally significantly smaller nuclear moments [3]). The interaction potential arises from two sources: the inherent spin of the electron; and the orbital motion of the electron around the nucleus. In some cases, such as the transition metals, the orbital part of the interaction is effectively quenched by the crystal structure.

In simplifying the derivation of the paramagnetic differential scattering cross section it is common to assume LS coupling [3, 48]. This assumption states that for individual electrons of spin **s** and orbital angular momentum **l**:

$$\mathbf{S} = \sum_i \mathbf{s}_i \quad \mathbf{L} = \sum_i \mathbf{l}_i \tag{3.25}$$

where **S** and **L** are the total spin and total orbital angular momentum respectively. However despite the tendency to employ LS coupling, it is possible that in certain cases, such as heavy atoms, it would be more accurate to employ jj-coupling, which allows for single electron spin orbit coupling [4]. Fortunately in the case of elastic neutron scattering the resulting difference in magnetic form factors due to using LS coupling instead of jj-coupling is minimal [4].

In the case of a free ion the paramagnetic differential scattering cross section is [4]:

$$\left(\frac{d\sigma}{d\Omega}\right)_{mag} = \left(\frac{\gamma e^2}{m_e c^2}\right)^2 \frac{1}{6} J(J+1) g^2 F^2(q) \tag{3.26}$$

where F(q) is the magnetic form factor, commonly expressed as [50]:

$$\begin{aligned} F^2(q) = {} & \langle j_0(q)\rangle^2 + C_{02}\langle j_0(q)\rangle\langle j_2(q)\rangle + C_{22}\langle j_2(q)\rangle^2 + C_{24}\langle j_2(q)\rangle\langle j_4(q)\rangle \\ & + C_{44}\langle j_4(q)\rangle^2 + C_{46}\langle j_4(q)\rangle\langle j_6(q)\rangle + C_{66}\langle j_6(q)\rangle^2 \end{aligned} \tag{3.27}$$

where C_{nm} are coefficients (given in Balcar and Lovesey [3]) and $\langle j_n(q)\rangle$ are functions dependent on spherical Bessel functions, $j_n(qr)$, and the normalised density of unpaired electrons, $f(r)$:

$$\langle j_n\rangle = \int_0^\infty r^2 \left[f(r)\right]^2 j_n(qr)\, dr \tag{3.28}$$

These functions have been calculated from first principles by several authors [25, 26] for both the transition metals and the rare earths. Using these values and Eqs. (3.26) and (3.27) it is possible to calculate the paramagnetic contribution to the differential scattering cross section and remove it.

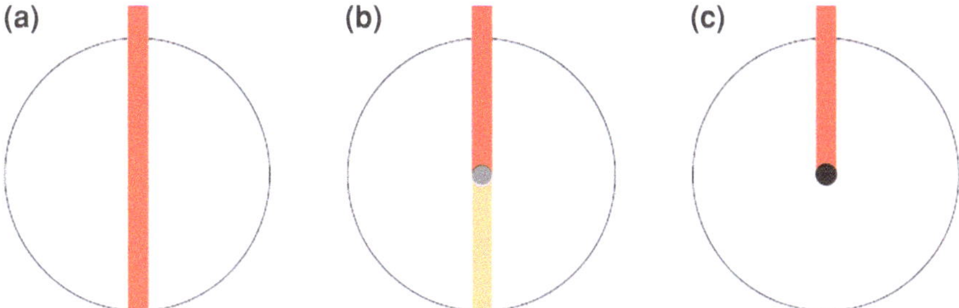

Fig. 3.3 A diagram illustrating the correction to the background required for sample self shielding. **a** The background intensity is measured. **b** A partially absorbing sample causes the background due to anything after the beam to be reduced. **c** The background intensity measured with a completely absorbing sample allows for correction of the effective decrease in the background in **a** that the sample experiences in **b**

3.2.7 Data Correction

Having discussed the general theory behind neutron scattering from amorphous structures, it is important to mention the essential considerations in correcting and normalising the detected intensity to give the differential scattering cross section. There are essentially five corrections that need to be applied in the case of neutron diffraction: paramagnetic scattering as discussed in Sect. 3.2.6; correcting the background subtraction for sample self shielding [5] removing the contribution of multiple scattering [7]; accounting for the effects of inelasticity [37]; and correcting the data for sample attenuation [36]. Small summaries of these corrections are given below.

3.2.7.1 Sample Self Shielding

In order to determine the sample scattering from a diffraction measurement it is important to remove the ambient scattering contributed purely from the instrument and its surroundings. The common method for this is to measure the scattering from the diffractometer without the sample. However this will overestimate the background at small angles (generally considered [5] to be <20°) caused by anything 'down beam' from the sample, as the sample absorption will effectively reduce this background during a sample run. This is most evident in diagrammatic form, as shown in Fig. 3.3. In order to correct for this, data must be collected from a totally absorbing sample (such as Cd or ^{10}B) of the same dimensions as the measured sample. This then allows the small angle background to be set to a value between the empty diffractometer and the total absorber, which is dependent on the attenuation of the sample:

$$I_S^B(\theta) = I_{Ab}^B(\theta) + A_{i,j}(\theta)\left[I_0^B(\theta) - I_{Ab}^B(\theta)\right] \tag{3.29}$$

where I_S^B is the corrected background to be removed from the sample, I_{Ab}^B is the background with an absorbing sample (Fig. 3.2c), I_0^B is empty background (Fig. 3.2a), and A_{ij} are the Paalman and Pings attenuation coefficients [36] to be discussed in Sect. 3.2.7.4. From this it is obvious that this correction is only significant in the case of strongly absorbing samples.

3.2.7.2 Multiple Scattering

In attempting to correct for multiple scattering, a significant problem lies in fact that an exact correction would require complete knowledge of the structure and dynamics of the sample [47]. The simplest treatment follows the work of Blech and Averbach [7] and the assumption that the error in the structure factor this introduces will be insignificant if the amount of multiple scattering is less than 20 % of the scattered beam.

3.2.7.3 Inelastic Correction

The inelasticity correction, or Placzek correction as it is often known [37], compensates for the static assumption introduced earlier. As with the multiple scattering correction the derivation is beyond the scope of this work; however the important point is that the magnitude of the correction is minimal for heavy nuclei and high incident energies [48]. It should be mentioned that the Placzek correction is strongly dependent on the energy dependence of the detector efficiency.

3.2.7.4 Attenuation Correction

Unlike the multiple scattering and inelastic corrections which require full structural and dynamic information to calculate exactly, the attenuation correction can be evaluated exactly [47], assuming a solid sample. In the case of a cylindrical sample with an annulus, this attenuation correction, which is utilised in Chaps. 4 and 7, was determined by Paalman and Pings [36].

However in the case of aerodynamic levitation the samples are spherical, for which the attenuation correction, A_s, follows intuitively from the derivation of Paalman and Pings [36]:

$$A_s(2\theta) = \tfrac{2}{3}\pi R^3 \int_0^{0.5\pi}\int_0^{2\pi}\int_0^{R} r^2 \sin(\gamma) \exp\Big[-\mu\Big[\big[\alpha(r,\gamma) - \beta(r,\gamma,\phi)\big]^{\frac{1}{2}} + \varepsilon(r,\gamma,\phi,2\theta) + \big[\alpha(r,\gamma) - \beta(r,\gamma,\phi-2\theta)\big]^{\frac{1}{2}}\Big]\, dr\, d\phi\, d\gamma \tag{3.30}$$

where R is the sample radius, 2θ is the scattering angle, ϕ is the azimuthal angle, γ is the polar angle, and:

$$\alpha(r,\gamma) = R^2 - (r\sin(\gamma))^2 \quad \beta(r,\gamma,\phi) = (r\cos(\gamma)\sin(\phi))^2$$
$$\varepsilon(r,\gamma,\phi,2\theta) = r\cos(\gamma)[\cos(\phi) - \cos(\phi - 2\theta)]$$

Equivalent results for spherical sample geometries have recently been reported by Zeidler [55].

3.2.7.5 Vanadium Normalisation

In the case of neutron scattering there is also the opportunity to normalise the sample scattering without considering the instrument calibration, by measuring the scattering from vanadium [47]. As vanadium has a negligible coherent scattering cross section it is possible to normalise the sample scattering per atom, given the sample and vanadium volumes and number densities, and the incoherent scattering cross section of vanadium.

3.3 Extended Absorption X-ray Fine Structure

A good introduction to the theory of Extended Absorption X-ray Fine Structure (EXAFS) is given by Als-Nielsen and McMorrow [2], while a more detailed consideration of EXAFS and its application to amorphous materials is given by Filipponi [20, 21]. While a full discussion is beyond the scope of this thesis, a brief mention should be made of the salient points relevant to glassy samples.

In the case of a free atom, an X-ray photon with energy equal or in excess of a core electron binding energy results in promotion of an electron to the continuum; this causes a significant increase in the X-ray absorption, $\mu(E)$, termed an absorption edge (Fig. 3.4). Once above this energy the absorption decays smoothly [2]. However in the case of an atom with a local atomic environment, the interatomic potentials affect the final states available for the emitted photoelectron. As the absorption depends on the transition probability of the electron, this causes a variation in the absorption with energy which is experimentally observed as characteristic oscillations above the absorption edge. These oscillations are termed the EXAFS, and are commonly defined by Als-Nielson and McMorrow [2]:

$$\chi(E) = \frac{\mu(E) - \mu_0(E)}{\mu_0(E)} \tag{3.31}$$

where $\mu_0(E)$ is the absorption coefficient of the free atom.

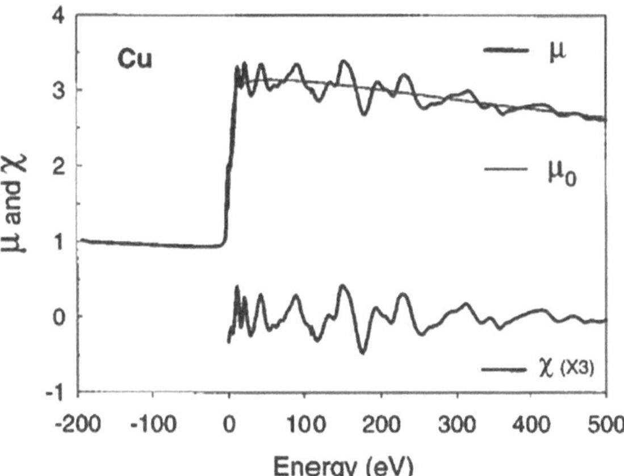

Fig. 3.4 The absorption edge structure of Cu illustrating the difference between the absorption in free atom case and the EXAFS case. Reproduced from Rehr and Albers [38] © 2000 by the American Physical Society

Assuming that the modification to the final state of the photoelectron, Δf, is small, it is possible to relate the EXAFS to the change in the absorption probability explicitly (Als-Nielson and McMorrow [2] as:

$$\chi(E) \propto \langle \Delta f | H | i \rangle \tag{3.32}$$

where i is the initial state and H is the interaction Hamiltonian. The change in the final photoelectron wavefunction due to the local environment can be interpreted as due to backscattering of the photoelectron. From this perspective it is possible to understand the components of the EXAFS equation [2]:

$$k\chi(k) = \sum_j N_j \frac{t_j(k)\,\sin\left(2kR_j + \delta_j(k)\right)}{R_j^2} e^{-2(k\sigma_j)^2} e^{-2R_j/\Lambda} \tag{3.33}$$

In this equation the backscattering is contributed from N_j atoms with a scattering amplitude $t_j(q)$ in a neighbouring shell j at a distance R_j; $\delta_j(q)$ is a phase shift resulting from the electrostatic potential of the surrounding ions; $e^{-2(q\sigma_j)^2}$ is a Debye-Waller term describing the vibrations of the neighbouring atoms; $e^{-2R_j/\Lambda}$ is a term that accounts for the possibility of inelastic scattering; and the energy dependence is instead expressed in terms of the photo-electron wavenumber, k.

From (3.33) it is a common task to extract the coordination numbers, N_j, and the nearest neighbour distances, R_j; however this is dependent on the ability to determine the scattering amplitude and phase shift. Programs such as FEFF [54] and GNXAS [22] are commonly used for these calculations.

3.3.1 Multiple Scattering Paths

In order to interpret the EXAFS from materials, it is necessary to describe the sum over multiple scattering paths interpretation of EXAFS [54]. This is particularly

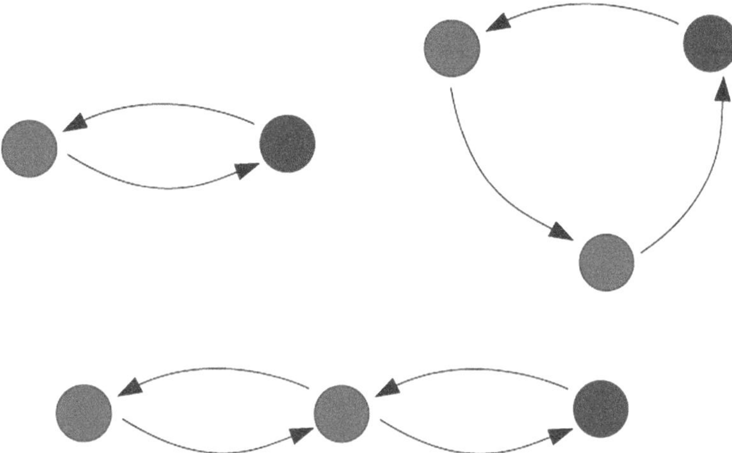

Fig. 3.5 An example of a single scattering path and multiple scattering paths with 3 and 4 legs

important when the target atom is in a highly ordered state. The backscattering described above can occur in two different ways: the primary mechanism is single-scattering, where the backscattering is just related to the path from the excited atom to the backscattering atom; and multiple scattering, where the backscattering occurs via more complex paths, such as those shown in Fig. 3.5. In order to calculate the theoretical scattering amplitude and phase shift, programs such as FEFF [54] calculate the multiple scattering paths based upon the atomic cluster defined. In the case of a crystal the atomic clusters are known; however they are ambiguous in disordered materials. Therefore in the case of an amorphous system where a number of different coordination environments are prevalent, the approach of fitting the theoretical EXAFS determined from the clusters based on the corresponding crystalline results in insufficient multiple scattering paths to accurately represent the real sample. However this is often mitigated by the smaller amount of multiple scattering due to the lower symmetry. Despite this, a different approach incorporating a sufficiently large cluster that all of the coordination environments are accounted for is occasionally required for a good fit to the EXAFS. If a cluster determined from an RMC configuration is used, the level of agreement can be used to validate the model.

3.4 Computer Simulations

3.4.1 Molecular Dynamics Simulations

Molecular dynamics (MD) simulations are an important tool for investigating atomic distributions in both liquids and glasses [1, 23]; however, because of their dependence on relatively simple empirical potentials, some degree of caution must be employed when interpreting their results. This dependence can be eliminated with the use of ab initio MD [12], which effectively continually calculates an approximation of the electronic structure. This allows the consideration of effects such as polarization, which are not described by simple empirical potentials, although can be incorporated through use of a polarizable-ion model [49]. However due to the significantly increased number of degrees of freedom present in ab initio MD, it requires an appreciable increase in computing time, which therefore limits simulations sizes to a maximum of ~1,000 atoms. As only MD simulations using empirical potentials were performed in the course of this work, this section will only discuss these in detail. All molecular dynamics simulations described in this thesis utilised the DL_POLY_2 molecular dynamics package [44].

The general principle of MD [1] is the numerical solution of Newton's laws of motion for a collection of particles with given initial position and velocity, and empirical potentials describing the interactions between particles. This is achieved by approximating the motion with a series of small timesteps; the smaller the timesteps the better the approximation. Although there are a number of different methods for calculating the future position, velocity and acceleration of a particle, the one employed in this work is the velocity Verlet algorithm [1], which works as follows:

Each atom starts with an initial position, **r**, velocity, **v**, and acceleration **a**, at time t. Given a simulation timestep δt, the velocity is calculated for half a timestep:

$$\mathbf{v}\left(t+\tfrac{1}{2}\delta t\right)=\mathbf{v}(t)+\tfrac{1}{2}\delta t\mathbf{a}(t) \tag{3.34}$$

This is then used to calculate **r**(t):

$$\mathbf{r}(t+\delta t)=\mathbf{r}(t)+\delta t\mathbf{v}\left(t+\tfrac{1}{2}\right) \tag{3.35}$$

which is effectively equivalent to calculating the new position using a Taylor expansion with three terms. This new position is then used to determine the force at $t+\delta t$, and therefore $\mathbf{a}(t+\delta t)$. The full time step velocity is then calculated using:

$$\mathbf{v}(t+\delta t)=\mathbf{v}\left(t+\tfrac{1}{2}\delta t\right)+\tfrac{1}{2}\delta t\mathbf{a}(t+\delta t) \tag{3.36}$$

Incorporating a half timestep velocity into the $\mathbf{r}(t+\delta t)$, $\mathbf{v}(t+\delta t)$, and $\mathbf{a}(t+\delta t)$ calculations effectively corrects the final velocity using the final acceleration. This algorithm has the advantage over a few others, such as the leapfrog Verlet [1], that at the end of each timestep **r**, **v**, and **a** are all known, which makes calculating thermodynamic quantities easier.

In practice the potentials are split into several different categories describing different types of interactions, such as two body chemical bonds, interactions related to valence and dihedral angles, and many body interactions. In this work only two body Buckingham potentials [10] were used:

$$U\left(r_{ij}\right)=A\exp\left(-\frac{r_{ij}}{\rho}\right)-\frac{C}{r_{ij}^{6}} \tag{3.37}$$

where r_{ij} is the distance between two particles i and j, and A, ρ, and C are pair specific constants. These potentials are determined [11] by fitting a Buckingham potential based model to experimental data of binary systems.

While it is possible to run MD simulations in a number of different ensembles, the MD simulations described in this thesis utilise an NVT ensemble. This allows the simulation density to remain equal to the experimental density [13].

3.4.2 Reverse Monte Carlo

One of the difficulties of evaluating one or more experimentally determined total pair correlation functions (or structure factors) is maximising the amount of structural information resulting from them. In cases where a coordination shell is distinct, it is possible to determine a nearest neighbour distance and a coordination number; however developing a full 3d structural model that fits the data to within statistical errors is difficult. One method for this is empirical potential structure refinement (EPSR) [45]. EPSR initially generates a configuration from Monte Carlo based on a specified reference potential function, before iteratively refining a further empirical potential with diffraction data. A limitation of EPSR is that as it uses a pairwise additive interaction potential, it cannot describe many-body effects, although it has been suggested that in certain circumstances three-body correlations can be achieved [46].

Another method, Reverse Monte Carlo (RMC) [33], is effectively a modification of the Metropolis Monte Carlo [34] (MMC) method; instead of specifying a pair potential and minimising the model's configurational energy, RMC requires one or more experimental pair correlation functions (or structure factors) and minimises the squared difference between the experimental $G(r)$ and the model $G(r)$. The RMC algorithm proceeds as follows:

1. An initial configuration of >1,000 atoms is required, this can be a random configuration, output of an MD simulation, or at stoichiometric compositions the equivalent crystal structure.
2. The equivalent function to the supplied experimental data is calculated based on this model, i.e. $F_{RMC}(q)$ if $F_{Exp}(q)$ is input. The squared difference between the two functions is calculated:

$$\chi^2 = \frac{(F_{RMC}(q) - F_{EXP}(q))^2}{\sigma(q)^2} \tag{3.38}$$

where $\sigma(q)$ is the experimental error.

3. A single atom is moved and the change in χ^2 is calculated. If χ^2 decreases the move is always accepted; however if it increases the move is accepted with probability:

$$P_{accept} = \exp\left(-\frac{(\Delta\chi^2)}{2}\right) \tag{3.39}$$

As with MMC it is essential that the RMC uses a Markov chain [1] for sampling states (so that the result is independent of the initial configuration) and that states are sampled ergodically. Assuming ergodicity this algorithm will result in fluctuations around the minimum of χ^2.

Although the initial configuration should have no influence upon the final structure, its selection is still important. Starting from an MD simulated configuration, even one determined by unsophisticated potentials, has significant advantages over a random configuration [28]. As the MD configuration should be closer to the final structure than a random configuration, the RMC should take less computational time to minimise χ^2. Another consideration is the final configuration determined by the RMC is not unique; it is just one possible configuration which results in a structure factor the same as the experimental structure factor, within experimental errors. Therefore if the RMC is effectively just a refinement of the failings of the empirical potentials, it increases the probability that the resultant configuration is similar to the real structure, without the need to include numerous constraints on the RMC. This also has the added benefit that, as there is a minimal distance in phase space between the starting configuration and the equilibrium configuration, the RMC is unlikely to get stuck in an unphysical local minimum. In order to ensure the MD configuration has only been refined, it is important to check the maximum atomic movements resulting from the RMC.

In this work experimental data is provided by neutron diffraction.

References

1. Allen MP, Tildesley DJ (1987) Computer simulations of liquids. Oxford University Press, Oxford
2. Als-Nielsen J, McMorrow D (2011) Elements of modern X-ray physics. Wiley, Singapore
3. Balcar E, Lovesey SW (1989) Theory of magnetic neutron and photon scattering. Oxford University Press, Oxford
4. Balcar E, Lovesey SW (1991) Theory of neutron scattering by atomic electrons: jj-coupling scheme. J Phys Condens Matter 3(37):7095–7115
5. Bertagnolli H, Chieux P, Zeidler MD (1976) A neutron-diffraction study of liquid acetonitrile. Mol Phys 32(3):759–773
6. Bhatia AB, Thornton DE (1970) Structural aspects of the electrical resistivity of binary alloys. Phys Rev B 2(8):3004–3012
7. Blech IA, Averbach BL (1965) Multiple scattering of neutrons in vanadium and copper. Phys Rev 137(4A):1113–1116
8. Bloch F (1936) On the magnetic scattering of neutrons. Phys Rev 50(3):259–260
9. Breit G, Wigner E (1936) Capture of slow neutrons. Phys Rev 49(7):519–532
10. Buckingham RA (1938) The classical equation of state of gaseous helium, neon and argon. Proc R Soc Lond A 168(933):264–283
11. Bush TS, Gale JD, Catlow RA, Battle PD (1994) Self-consistent interatomic potentials for the simulation of binary and ternary oxides. J Mater Chem 4(6):831–837
12. Car R, Parrinello M (1985) Unified approach for molecular dynamics and density-functional theory. Phys Rev Lett 55(22):2471–2474
13. Clark EB, Mead RN, Mountjoy G (2006) A molecular dynamics model of the atomic structure of Tb metaphosphate glass $(Tb_2O_3)_{0.25}$ $(P_2O_5)_{0.75}$. J Phys: Condens Matter 18(29):6815–6826

14. Containerless Research Inc. (1999) Conical nozzle levitation system preliminary operator's manual
15. Cossy C, Barnes AC, Enderby JE (1989) The hydration of Dy^{3+} and Yb^{3+} in aqueous solution: a neutron scattering first order difference study. J Chem Phys 90(6):3254–3260
16. Coté B, Massiot D, Taulelle F, Coutures J-P (1992) ^{27}Al NMR spectroscopy of aluminosilicate melts and glasses. Chem Geol 96(3–4):367–370
17. Dianoux A-J, Lander G (eds) (2003) Neutron data booklet, 2nd edn. Old City Publishing, Philadelphia
18. Enderby JE, North DM, Egelstaff PA (1966) The partial structure factors of liquid Cu-Sn. Phil Mag 14(131):961–970
19. Faber TE, Ziman JM (1965) A theory of the electrical properties of liquid metals. Phil Mag 11(109):153–173
20. Filipponi A (1994) The radial distribution function probed by X-ray absorption spectroscopy. J Phys: Condens Matter 6(41):8415–8427
21. Filipponi A (2001) EXAFS for liquids. J Phys: Condens Matter 13(7):R23–R60
22. Filipponi A, Cicco AD, Madonna V (1995) X-ray-absorption spectroscopy and n-body distribution functions in condensed matter 1 Theory. Phys Rev B 52(21):15122–15134
23. Fischer HE, Barnes AC, Salmon PS (2006) Neutron and x-ray diffraction studies of liquids and glasses. Rep Prog Phys 69(1):233–299
24. Fratello VJ, Brandle CD (1993) Physical properties of a $Y_3Al_5O_{12}$ melt. J Cryst Growth 128(1–4):1006–1010
25. Freeman AJ, Desclaux JP (1979) Dirac-Fock studies of some electronic properties of rare-earth ions. J Magn Magn Mater 12(1):11–21
26. Freeman AJ, Watson RE (1961) Hartree-Fock atomic scattering factors for the neutral atom iron transition series. Acta Crystallogr A 14(3):231–234
27. Hennet L, Thiaudière D, Landron C, Melin P, Price DL, Coutures J-P (2003) Melting behavior of levitated Y_2O_3. Appl Phys Lett 83(16):3305–3307
28. Kidkhunthod P, Barnes AC (2009) A structural study of praseodymium gallate glasses by combined neutron diffraction, molecular dynamics and EXAFS techniques. J Phys: Conf Ser 190:012076
29. King SM, Griffiths PC, Cosgrove T (2000) Using SANS to study adsorbed layers in colloidal dispersions (Chap. 4). In: Gabrys BJ (ed) Applications of neutron scattering to soft condensed matter. Gordon and Breach Science Publishers, Singapore
30. Krishnan S, Hennet L, Jahn S, Key TA, Madden PA, Saboungi M-L, Price DL (2005) Structure of normal and supercooled liquid aluminum oxide. Chem Mater 17(10):2662–2666
31. Krogh-Moe J (1966) A method for the resolution of composite radial pair distribution function. Acta Chem Scand 20(10):2890–2891
32. Landron C, Hennet L, Jenkins T, Greaves G, Coutures J-P, Soper A (2001) Liquid alumina: detailed atomic coordination determined from neutron diffraction data using empirical potential structure refinement. Phys Rev Lett 86(21):4839–4842
33. McGreevy RL, Pusztai L (1988) Reverse monte carlo simulation: a new technique for the determination of disordered structures. Mol Simul 1(6):359–367
34. Metropolis N, Rosenbluth AW, Rosenbluth MN, Teller AH, Teller E (1953) Equation of state calculations by fast computing machines. J Chem Phys 21(6):1087–1092
35. Nordine PC (1986) The accuracy of multi-color optical pyrometry. High Temp Sci 21(2):97–109
36. Paalman HH, Pings CJ (1962) Numerical evaluation of X-ray absorption Factors for cylindrical samples and annular samples cells. J Appl Phys 33(8):2635–2639
37. Placzek G (1952) The scattering of neutrons by systems of heavy nuclei. Phys Rev 86(3):377–388
38. Rehr JJ, Albers RC (2000) Theoretical approaches to X-ray absorption fine structure. Rev Mod Phys 72(3):621–654
39. Salmon PS (1992) The structure of molten and glassy 2:1 binary systems: an approach using the Bhatia-Thornton formalism. Proc R Soc Lond A 437(1901):591–606

40. Sivia DS (2011) Elementary scattering theory. Oxford University Press, Oxford
41. Skinner LB (2008) Structure and phase behaviour of Aluminate glasses, produced using levitation and laser heating. University of Bristol, Ph. D
42. Skinner L, Barnes AC (2006) An oscillating coil system for contactless electrical conductivity measurements of aerodynamically levitated melts. Rev Sci Instrum 77(12):123904
43. Skinner LB, Barnes AC, Salmon PS, Crichton WA (2008) Phase separation, crystallization and polyamorphism in the Y_2O_3-Al_2O_3 system. J Phys: Condens Matter 20:205103
44. Smith W, Forester T (1996) DL POLY 2.0: A general-purpose parallel molecular dynamics simulation package. J Mol Graph 14(3):136–141
45. Soper AK (1996) Empirical potential Monte Carlo simulation of fluid structure. Chem Phys 202(2–3):295–306
46. Soper AK (2001) Tests of the empirical potential refinement method and a new method of application to neutron diffraction data on water. Mol Phys 99(17):1503–1516
47. Soper AK, Howells WS, Hannon AC (1989) ATLAS—analysis of time-of-flight diffraction data from liquid and amorphous samples. Rutherfod Appleton Laboratory report RAL-89-046
48. Squires GL (1978) Introduction to the theory of thermal neutron scattering. Cambridge University Press, Cambridge
49. Takagi R, Hutchinson F, Madden PA, Adya AK, Gaune-Escard M (1999) The structure of molten $DyCl_3$ and $DyNa_3Cl_6$ simulated with polarizable- and rigid-ion models. J Phys: Condens Matter 11(3):645–658
50. Waase JC, Salmon PS (1999) Structure of molten lanthanum and cerium tri-halides by the method ofisomorphic substitution in neutron diffraction. J Phys: Condens Matter 11(6):1381–1396
51. Westlake JR (1968) A handbook of numerical matrix inversion and solution of linear equations. Wiley, New York
52. Wille G, Millot F, Rifflet JC (2002) Thermophysical properties of containerless liquid iron up to 2,500 K. Int J Thermophys 23(5):1197–1206
53. Wright AC, Etherington G, Erwin Desa JA, Sinclair RN (1982) Neutron diffraction studies of rare earth ions in glasses. J de Phys 43(9):31–34
54. Zabinsky SI, Rehr JJ, Ankudinov A, Albers RC, Eller MJ (1995) Multiple-scattering calculations of x-ray absorption spectra. Phys Rev B 52(4):2995–3009
55. Zeidler A (2012) X-ray and neutron attenuation correction factors for spherical samples. J Appl Crystallogr 45(1):122–123

Chapter 4
Rare Earth Doped Barium Titanate Glass

4.1 Introduction

$BaTi_2O_5$ (hereafter referred to as BTO), which was first reported relatively recently [34], is of interest from both technological and fundamental perspectives.

Crystalline and ceramic $BaTiO_3$ are archetypal ferroelectric materials that have been well known for a long time [15, 20]. By contrast ferroelectricity was only discovered in crystalline BTO recently [1]. The ferroelectric properties of BTO appear to be more technologically favourable than $BaTiO_3$, as it possesses both a higher dielectric constant and a ferroelectric T_C 300 K higher [2]. Although ferroelectricity is generally associated with small, invertible distortions of a high symmetry crystal structure, it has been suggested that a ferroelectric glass is not fundamentally forbidden [14]. This has led to interest in glasses produced from ferroelectric crystals. Although there is some ambiguity in the literature with ferroelectric glass occasionally being used to describe a glass produced from a ferroelectric crystal [34], I am unaware of any demonstration of an amorphous material which possess a spontaneous electric polarization.

Alongside the possibility of producing a pure ferroelectric glass is the controlled nucleation of crystals for ferroelectric glass–ceramics [4, 18]. This has been shown [33] to improve the ferroelectric properties of $BaTiO_3$ microcrystals in a sol–gel derived borosilicate, relative to pure $BaTiO_3$ ceramic.

The doping of BTO with rare earth ions up to compositions of $RE_{0.3}Ba_{0.7}Ti_2O_{5.15}$ has been demonstrated by [35]. The addition of Er^{3+} shows an increase in both T_g and refractive index with concentration, with $T_g = 750$ °C and $n_d = 2.22$ (at 579 nm) at maximum doping [16]. Masuno et al. [16] have also reported that the phonon energy in BTO glass is lower than in oxide and tellurite glasses. This is an important property, as the decay of rare earth fluorescence has been shown [31, 32] to be due multiphonon quenching. This multiphonon quenching occurs due to clustering of the rare earth atoms, which Arai et al. [3] reported is due to rare earth sharing NBO in the case of rare earth doped aluminosilicates.

T. Farmer, *Structural Studies of Liquids and Glasses Using Aerodynamic Levitation*, Springer Theses, DOI: 10.1007/978-3-319-06575-5_4,
© Springer International Publishing Switzerland 2015

Aside from the technological considerations, BTO glass is also interesting from the fundamental perspective of glass formation, as it does not contain an archetypal good glass former [29]. Although in this respect BTO glass is similar to aluminate glasses [26, 27], such as $BaAl_3O_{5.5}$, unlike the Al in aluminate glasses, Ti atoms are often found to exhibit fivefold (or greater) coordination in other glasses [12]. If the glass formation of atypical glass formers is to be understood it is important that these underlying network differences are well described.

4.1.1 Crystal Structure

Stable crystalline BTO is monoclinic [1] with unit cell parameters: $a = 16.91$ Å, $b = 3.94$ Å, $c = 9.49$ Å and $\beta = 103.0°$. This corresponds to a Ti–O coordination of 6 at distance of 1.76–2.47 Å, and a Ba–O coordination of 12 at a distance of 2.67–3.28 Å. It has also been shown [35] to exist in two metastable forms, α and β, where only the structure of β has been determined (Pnma space group with $a = 10.2$ Å, $b = 3.93$ Å, $c = 10.9$ Å).

4.2 Methods

Samples of BTO and $RE_{0.3}Ba_{0.7}Ti_2O_{5.15}$ (where RE = Nd, Gd, Er, Yb, hereafter referred to as RE–BTO) glass were formed from $BaTiO_3$ (99.995 % pure), TiO_2 (99.99 % pure) and RE_2O_3 (min 99.99 % pure) powders using aerodynamic levitation. All samples were levitated on Ar gas in an oxidising atmosphere and quenched at the maximum quench rate achieved by instantaneous laser shutoff. Samples were checked for hints of crystallization using both optical microscopy and pyrometric studies, as described in Sect. 3.1.1. Samples prepared for XAS were produced using the multi ball method described in Sect. 3.1.2. Due to the much larger number of samples required for neutron diffraction, in this case samples were produced by another method; initially a bulk mixture of the correct proportions was sintered in a furnace at 1,000 °C, before being ground into homogenous powder using a pestle and mortar. The powder was then fused on a copper hearth and levitated. Samples were compositionally accurate to within the uncertainty in the EDX (~1 %).

Neutron diffraction was performed on the D4c diffractometer [10] at the ILL at a wavelength of 0.4972 Å. Glass samples of the same composition, for example all BTO samples, were loaded into a 5 mm inner diameter vanadium can. The total mass of samples at each composition is shown in Table 4.1. The average glass sample radius was ~1.7 mm for the pure BTO and ~1.4 mm for the doped samples. An empty vanadium can, boron powder in a vanadium can, the empty belljar, and a nickel standard were also measured for calibration and correction purposes. All measurements were made at room temperature. A sample packing fraction of 0.51 ± 0.05 was determined from liquid volume measurements. All samples were corrected for

Table 4.1 The total mass of spherical samples placed in a vanadium can for neutron scattering

Composition	Total mass (mg)
BTO	171
Nd–BTO	196
Gd–BTO	391
Er–BTO	183
Yb–BTO	60

Table 4.2 The resonance parameters for the three absorption resonances at thermal neutron wavelengths for ^{155}Gd, and the two absorption resonances at thermal neutron wavelengths for ^{157}Gd [23]

Isotope	Abundance (%)	b_{real} (fm) (1.798 Å)	b_{imag} (fm) (1.798 Å)	E_j (eV)	Γ_n^0 (meV)	Γ (eV)
^{155}Gd	14.8	13.80	16.99	0.0268	0.397	0.108
				2.008	0.098	0.11
				2.568	0.68	0.111
^{157}Gd	15.7	4.000	72.02	0.0314	1.65	0.106
				2.825	0.128	0.097

Table 4.3 The established edge energies for edges probed in the XAS measurements [30]

Element and edge	Edge energy (keV)
Ti K-edge	4.966
Nd L_3-edge	6.208
Gd L_3-edge	7.243
Er L_3-edge	8.358
Yb L_3-edge	8.944

paramagnetic scattering due to their rare earth components, and followed the standard corrections described in Sect. 3.2.7. The Gd doped samples were also corrected for the neutron resonance contributions from two isotopes, ^{155}Gd and ^{157}Gd, as explained in Sect. 3.2.4 with the relevant parameters shown in Table 4.2.

XAS measurements were made in fluorescence mode on the B18 core EXAFS beamline [6] at the Diamond synchrotron. An Si(111) monochromator with energy resolution of 0.014 %, and a beam height of 250 μm and beam width of 200 μm were used to measure coordinations around the edges shown in Table 4.3. Due to the overlap of the Ti K-edge it was not possible to accurately measure the fluorescence from the Ba L-edges, which lie just above 5 keV. In order to mount the glasses, single samples of each composition of between 1.5 and 2.5 mm diameter were set in epoxy resin cylinders, which was then polished down to expose the sample surface. Each cylinder was then positioned in the X-ray sample holder at a 45° angle to both the beam and the Ge solid state detector (Fig. 4.1). A Ti foil (K-edge energy 4,966 eV) was measured for energy calibration. Due to time constraints the Yb doped BTO was not measured at the Ti K-edge.

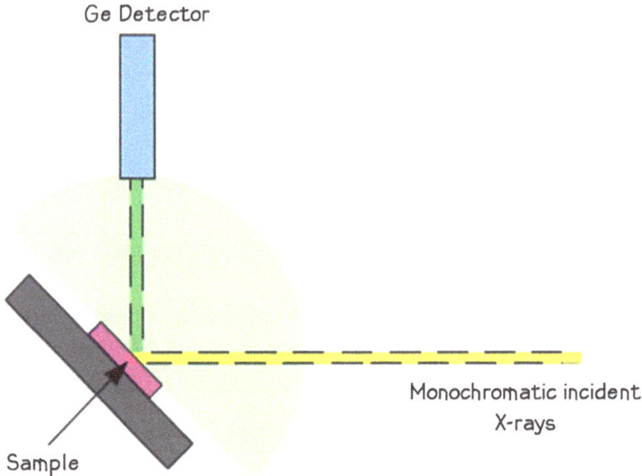

Fig. 4.1 A schematic of the detector position for XAS fluorescence measurements

Table 4.4 Formal ion charges were used in all cases

	A (eV)	ρ (Å)	C (eV Å^{-6})
Ba^{2+}–O^{2-}	4818.416	0.3067	0
Ti^{4+}–O^{2-}	2088.107	0.2888	0
O^{2}–O^{2-}	25.41	0.6937	32.32
Nd^{3+}–O^{2-}	13084.2170	0.255	0
Gd^{3+}–O^{2-}	866.339	0.3770	0
Y^{3+}–O^{2-}	1519.279	0.3291	0
Yb^{3+}–O^{2-}	1309.6	0.3462	0

	A	B	σ	C	D
BHM fit O^{2-}–O^{2-}	12.604	1.441	0.498	41.329	−41.52

In order to correct for the low r attractive behaviour of the O^{2-}–O^{2-} Buckingham potential, a BHM type potential was fitted to it. This differed from the Buckingham potential by less than 5 % at 2 Å < r < 6.7 Å. All potentials were taken from Bush and Catlow [5], except the Yb^{3+}–O^{2-} potential which was taken from Lewis and Catlow [13]. In the absence of an Er^{3+}–O^{2-} potential the Y^{3-}–O^{2-} potential was used, as Y and Er are isomorphs

Molecular dynamics simulations were performed using the DL_POLY_2 [28] molecular dynamics package and the Buckingham potentials shown in Table 4.4. Due to the C term in the O–O potential, the short range behaviour becomes attractive below 1.5 Å. This was corrected by fitting the O–O potential with a Born–Huggins–Mayer type potential [11]:

$$U(r_{ij}) = A\exp\left[B(\sigma - r_{ij})\right] - \frac{C}{r_{ij}^6} - \frac{D}{r_{ij}^8} \tag{4.1}$$

Table 4.5 The closest approaches used in the MD and the RMC. The RE–RE, RE–Ti and RE–O closest approaches were changed in the RMC to the bracketed numbers, in order to exclude unphysical spikes that arose at shorter distances

Pair	Closest approach (Å)
RE–RE	2.5 (2.7)
RE–Ba	2
RE–Ti	2 (2.7)
RE–O	1.6 (1.9)
Ba–Ba	3
Ba–Ti	3
Ba–O	2
Ti–Ti	2.5
Ti–O	1.6
O–O	2.4

Table 4.6 The RMS distance of atomic movement during the RMC refinement

Sample	RMS Ti (Å)	RMS O (Å)	RMS RE (Å)	Average (Å)
BTO	0.78	0.99	–	0.93
Nd BTO	0.56	0.78	0.98	0.74
Gd BTO	0.62	0.85	1.06	0.80
Er BTO	0.52	0.77	1.44	0.73
Yb BTO	0.73	0.90	0.80	0.85

The Ba atoms were not included in the calculations as these were not allowed to move during the RMC. The slightly lower average movement distance for the rare earth glass RMCs should be expected as there are a larger number of different contributions to the $S(q)$ i.e. more degrees of freedom. The larger RMS movement of the Er atoms is probably due to the use of the Y^{3+}–O^{2-} potential

where r_{ij} is the atomic separation, and A, B, σ, C, and D are empirical parameters. The fit differed from the Buckingham potential by less than 5 % over the range 2 Å $< r <$ 6.7 Å, by which distance the potential is $\approx$0.

An Ewald sum acting up to 12 Å was employed to account for long range electrostatic interactions. Each initial configuration was randomly generated and then restricted with the closest approaches shown in Table 4.5, which were underestimates based on the partials pair correlation functions of Yu et al. [35] In the case of BTO 4,800 atoms and a number density of 0.0695 was used, resulting in a box size of ~41 Å. For the rare earth doped glasses, 4,890 atoms and a number density of 0.07 was used, again resulting in a box size of ~41 Å. All simulations were run on a desktop computer for a minimum of 20 h to ensure an equilibrium, energy minimised structure resulted.

The output configurations generated by the DL_POLY_2 MD [28] were used as the input configurations for the RMC program RMCA (version 3.14) [19]. Starting from this configuration the simulation rapidly converged, with the RMS atomic movements shown in Table 4.6. The same closest approaches were used to constrain the RMC, except in the cases of the rare earth closest distances (Table 4.5). The RMC was fitted to the neutron $F(q)$, as shown in Fig. 4.10a. For each sample

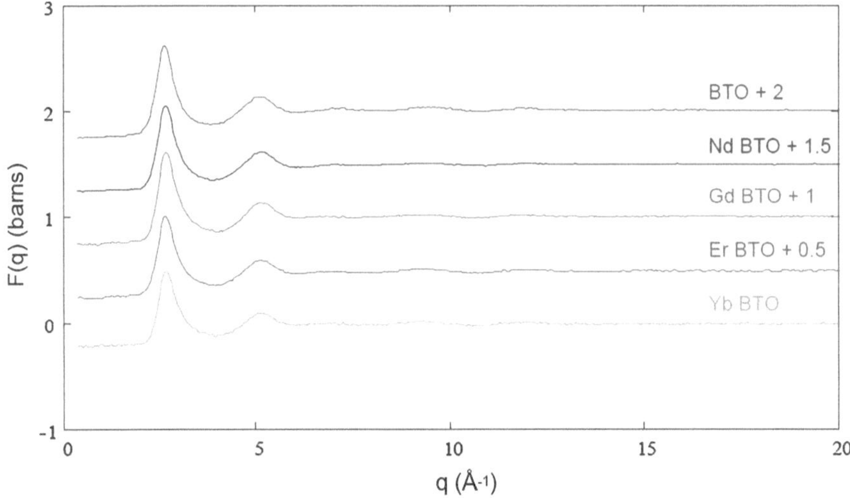

Fig. 4.2 The $F(q)$'s for BTO glass and the rare earth doped BTO glasses. As might be expected from the relatively low doping (~3.5 %) there is no significant difference between the pure BTO glass and the doped BTO glasses. The noise in the data is indicative of the error

the Ba atom positions were fixed in the RMC, as otherwise the Ba–Ba partial pair correlation function developed an unphysically sharp rise above the closest approach distance, similar to those caused by RMC starting from a random configuration. This occurs because of the low weighting of Ba in the neutron diffraction measurements and the overlap of correlation functions. The acceptance to rejection ratios for all of the RMC simulations lay in the region of 0.36–0.4.

4.3 Results

The total structure factors and total pair correlation functions from the neutron diffraction are shown in Figs. 4.2 and 4.3a. There is a high degree of similarity in the total structure factors for the five samples. However in the total pair correlation functions there appears to be a degree of difference between the five samples occurring at a distance of ~2.3 Å.

Figure 4.3b shows the comparison between the original $S(q)$'s and the back Fourier transforms of the low r cutoff $G(r)$'s, in order to check for consistency. It is important to note that there is a small amount of difference between the two at low q in the case of the Gd and Yb doped BTO glass. The cause of this inconsistency is yet to be determined, although it seems probable that in the case of the Gd it is related to the absorption resonances in ^{155}Gd and ^{157}Gd. A significant difficulty in the Gd BTO analysis is checking for consistency with the high q limit, as Gd has a very high incoherent scattering cross section (151b at an incident neutron wavelength of 1.798 Å [7]). Calculating the contribution of this incoherent scattering to the high

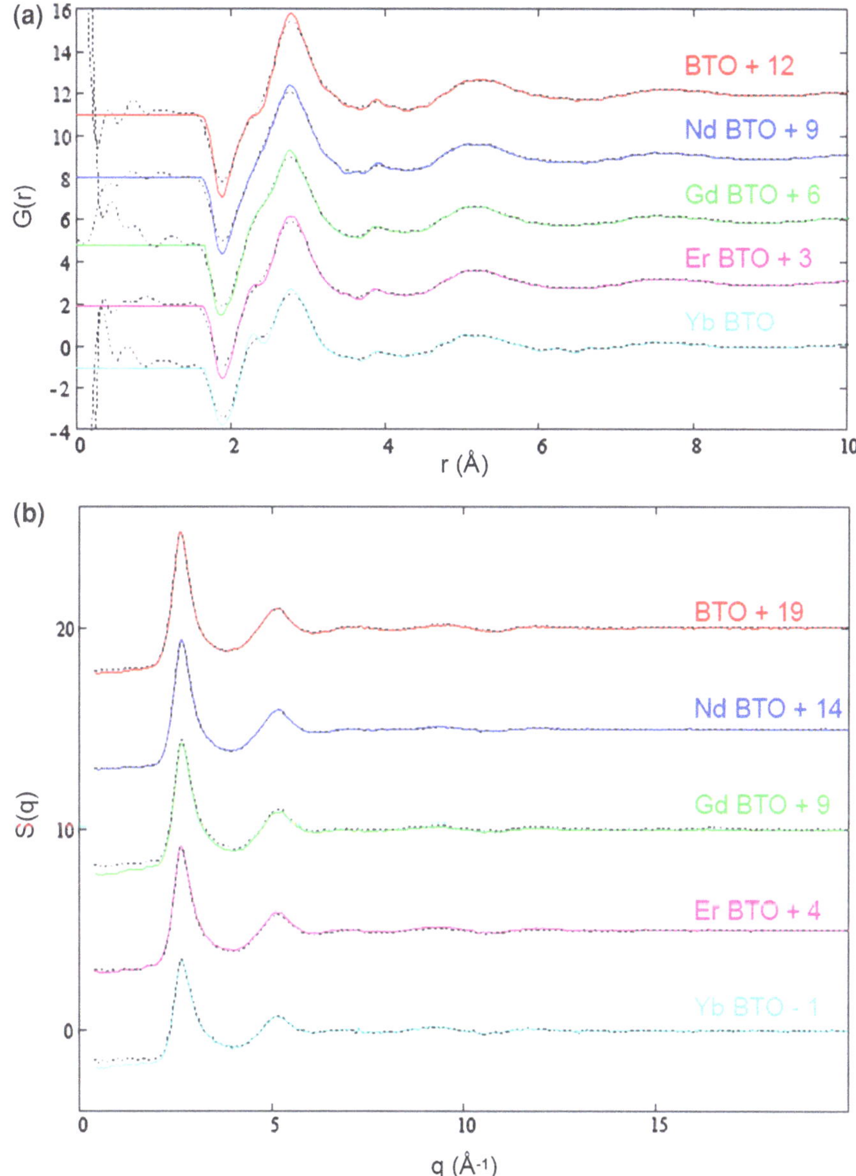

Fig. 4.3 **a** The prefactor normalised $G(r)$'s for BTO glass and the rare earth doped BTO glasses. The *coloured solid lines* show the $G(r)$'s with the low r cutoffs, while the *black dotted lines* show the Lorch modified $G(r)$'s. **b** The *coloured solid lines* are the back Fourier transformed $S(q)$'s from the cutoff $G(r)$'s from **a**, and the *dotted black lines* are the original $S(q)$'s

q limit given that it also varies with wavelength due to the absorption resonances, results in a limit to the reliability of this consistency check.

The Ti K-edge and RE L_3-edge EXAFS results are shown in Figs. 4.4 and 4.5. The normalisation on the absorption was determined by fitting a line to the

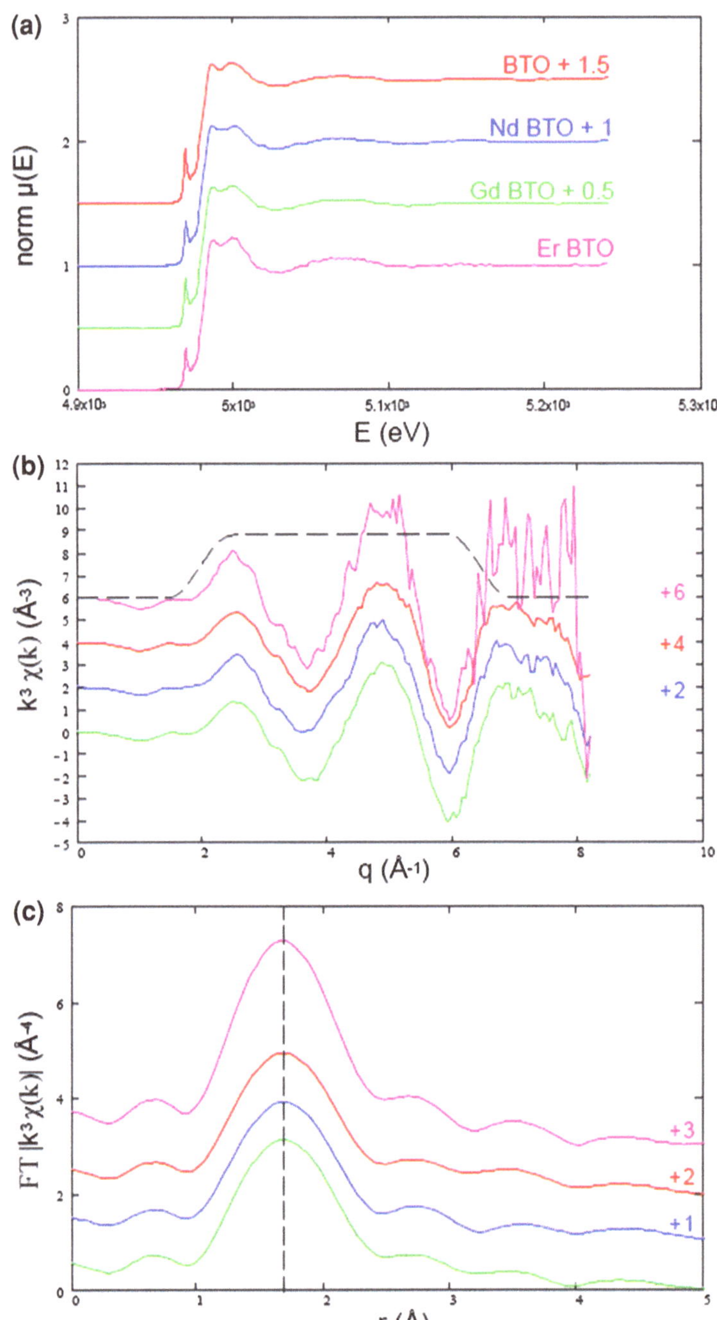

Fig. 4.4 The Ti K-edge EXAFS shown as **a** the normalised absorption against energy, **b**$k^3\chi(k)$ against q, and c) $\chi(r)$ against r. The *black broken line* in **b** shows the Hanning window function applied for the Fourier transform to real space. The *vertical grey broken line* in **c** illustrates the constancy of the peak position. In all diagrams: BTO glass (*red*), Nd doped BTO (*blue*), Gd doped BTO (*green*), and Er doped BTO (*pink*). Traces have been shifted up for clarity as indicated

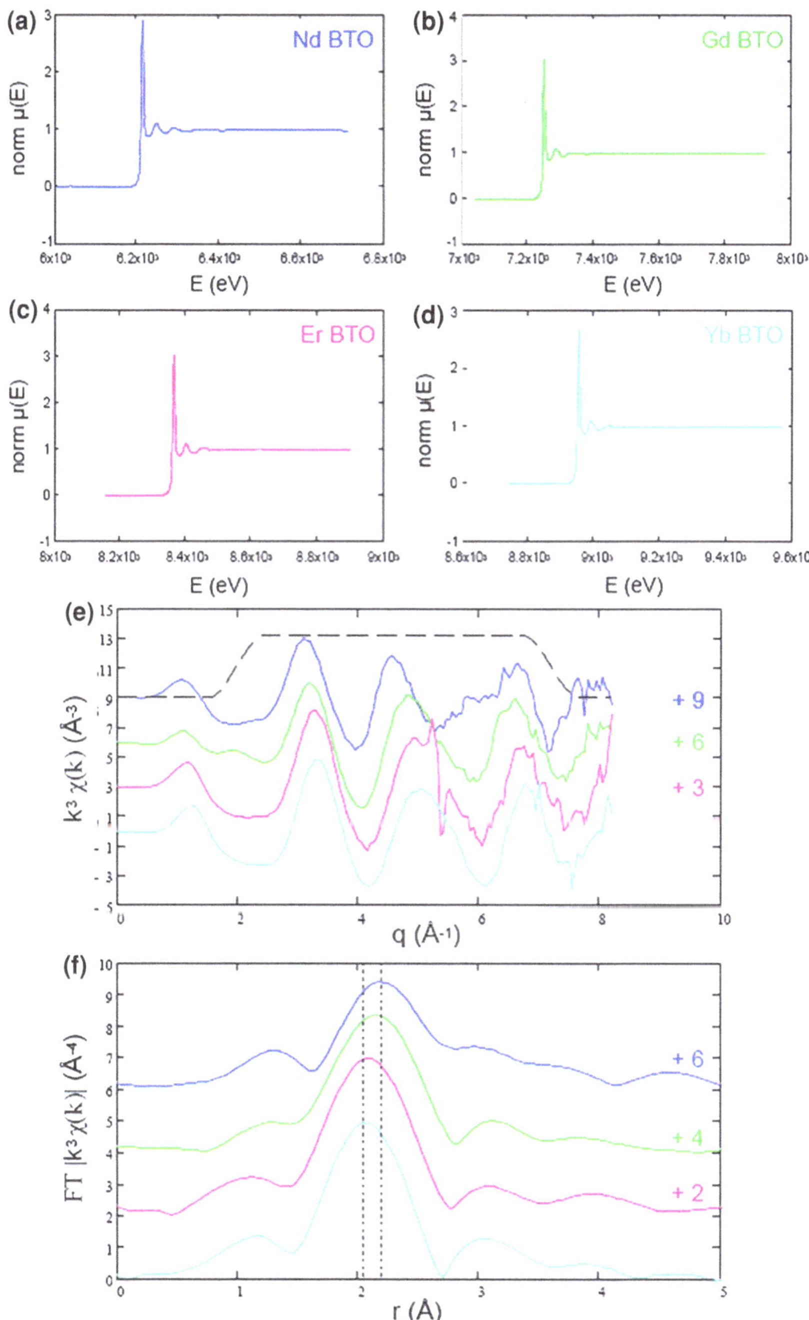

Fig. 4.5 The RE L_3-edge EXAFS shown as the normalised absorption against energy for **a** Nd doped BTO (*blue* in all diagrams), **b** Gd doped BTO (*green*), **c** Er doped BTO (*pink*), and **d** Yb doped BTO (*light blue*), and for all samples **e** $k^3\chi(k)$ against q, and **f** $\chi(r)$ against r. The *black broken line* in **e** shows the Hanning window function applied for the Fourier transform to real space. The *vertical black dotted lines* in **c** illustrate the change in the peak position of ~0.12 Å

Table 4.7 The prepeak heights relative to the first peak height after the absorption edge, and the prepeak positions relative to the first inflection point of a Ti foil standard

Sample	Relative energy (eV)	Relative height
BTO	4	0.34
Nd BTO	4.2	0.27
Gd BTO	4	0.30
Er BTO	4	0.23

The errors on the relative energy were ±0.98 eV, due to the beamline resolution. The errors on the relative height were ±0.01. The Er BTO XAS data is significantly noisier than the others, which resulted in difficulties when determining the edge step

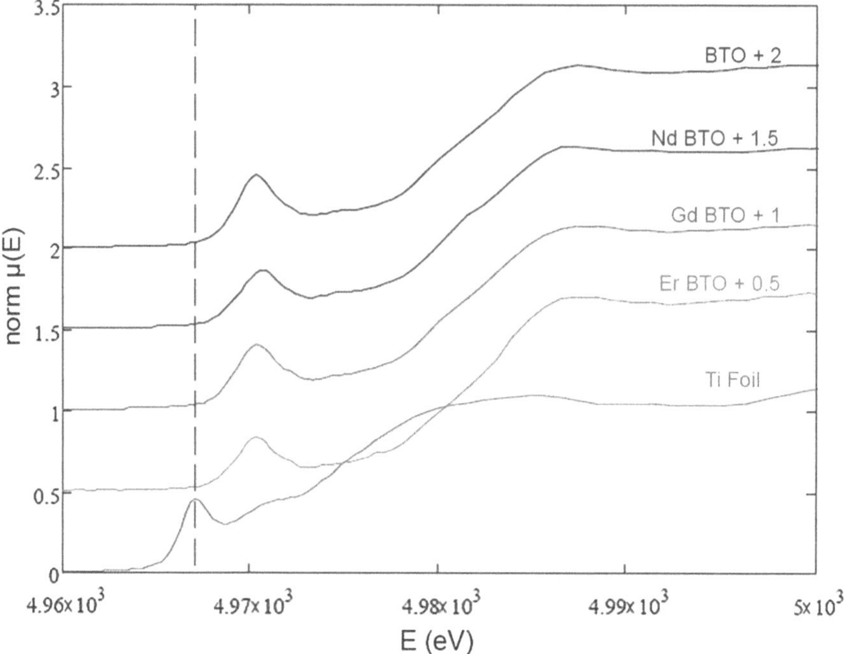

Fig. 4.6 The XANES region of the Ti–K edge measurements for BTO (*red*), Nd (*blue*), Gd (*green*), and Er (*pink*). The Ti foil which was used for calibration is shown in *grey* with a *broken black line* showing the energy of the first peak

pre-edge region and a quadratic to the post-edge region using the data treatment program Athena, part of the IFEFFIT EXAFS analysis suite [24]. The background correction was performed using the rbkg routine which subtracts a spline fit above the edge energy, and a Hanning type window was fitted for Fourier transform to real space. From the similarity of the Ti-edge EXAFS data for all samples, it appears there is little change in the local environment around the Ti atom with either the introduction or variation of rare earth dopants. However in the case of the RE-edge data there is a significant difference, which can be seen in the real space data to be related to a reduction in the nearest neighbour distance

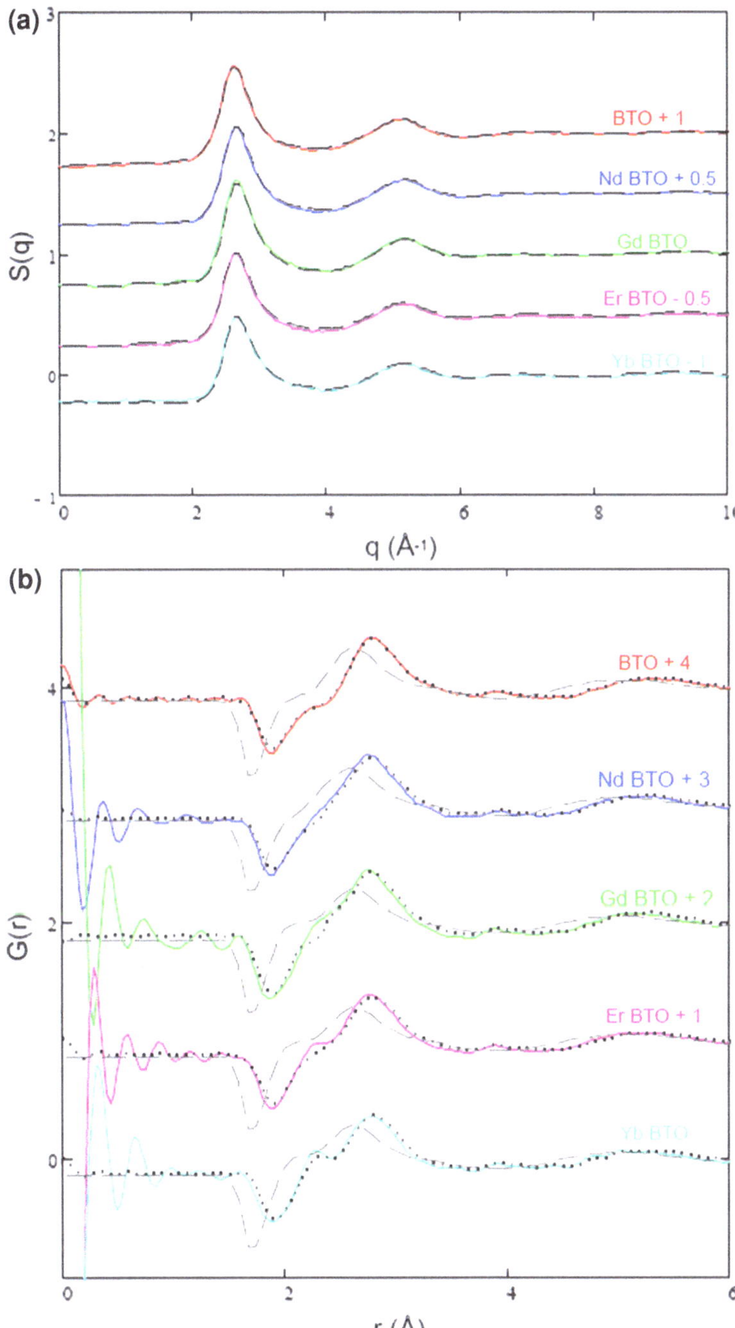

Fig. 4.7 **a** The neutron diffraction total structure factors (*solid coloured lines*) compared with the RMC fits (*broken black lines*). **b** The neutron diffraction total pair correlation functions (*solid coloured lines*) compared with those determined from the initial MD (*broken grey lines*) and the RMC (*dotted black lines*). While the MD has similar features to the diffraction data, the distances are reduced by ~0.2 Å

with increasing rare earth atomic mass. As mentioned in Sect. 3.3, multiple coordination environments make fitting to EXAFS data challenging. Initial attempts were made to fit the *k*-space data using scattering paths generated from the crystal structures of BTO and $BaTiO_3$, although these were unsuccessful. Ideally it would be possible to directly calculate the EXAFS spectra from the RMC configuration, which could then be used for fitting; however this is currently not possible in RMCA and its derivatives.

Along with the EXAFS, some consideration should also be given to the XANES region of the spectra (Fig. 4.6). Following the work of Farges et al. [8, 9, 22], the prepeak region (from 4,940 to 4,980 eV) was fitted with Lorentzians. In each sample the main XANES peak fit (centred on ~4,970 eV) was almost completely contributed by a single Lorentzian. The height of this peak and position relative to the Ti foil first peak is compared in Table 4.7. However due to the absence of measured 4, 5 and 6 coordinated standards, and the lack of calibration measurements taken simultaneously with sample measurements, the analysis of the XANES region is limited (Fig. 4.6).

The total pair correlation functions for the neutron diffraction, MD, and RMC of the all samples are compared in Fig. 4.7. While there is broad similarity between the MD and the neutron data, it is apparent that there are significant interatomic interactions that are not accounted for with the simple Buckingham potentials described in Sect. 4.2. The RMC total structure factors are in very good agreement with the corresponding neutron diffraction structure factors, as shown in Fig. 4.7a.

4.4 Discussion

The results of the pure BTO glass can be compared with the work of [35], who used neutron and X-ray diffraction to constrain a random starting configuration RMC. The neutron structure factors are in good agreement with respect to both peak positions and heights. Similarly, the partial pair correlation function determined by the RMC shows generally comparable behaviour, although it is important to note the influence of the MD starting configuration on the sharpness of the partials. The one exception is in the Ti–Ti partial pair correlation function. While a bifurcation of the first peak is visible in both RMCs, there is a significant difference in peak height. In most cases the fact that the RMC of Yu et al. [35] is constrained by two experimental datasets would suggest the partial pair correlation functions would be more reliable; however the Ti–Ti first peak is located at a distance also occupied by the Ba–O and O–O first peaks. The Ba–O has a similar magnitude contribution in the X-ray data as the Ti–Ti, which is also swamped in the neutron data; this results in the RMC being relatively insensitive to the Ti–Ti partial.

Due to the corresponding crystalline bond lengths [1] and the negative scattering of Ti, it is reasonable to assume that the first peak in the total correlation

Table 4.8 The Ti–O coordination numbers for all samples determined from neutron diffraction

Sample	Running	Gaussian	Reflected peak
BTO	4.7 ± 0.1 (2.32 Å)	4.7 ± 0.1	3.69 ± 0.09
Nd BTO	4.9 ± 0.1 (2.24 Å)	5.0 ± 0.1	3.93 ± 0.09
Gd BTO	5.4 ± 0.1 (2.20 Å)	5.6 ± 0.1	4.1 ± 0.1
Er BTO	4.4 ± 0.1 (2.18 Å)	4.5 ± 0.1	3.92 ± 0.09
Yb BTO	3.86 ± 0.08 (2.15 Å)	4.03 ± 0.09	3.33 ± 0.07

An example of the two Gaussian fit is shown in Fig. 4.8. The running coordination numbers were calculated based up to an r value where $G(r)$ became positive. These values are shown in brackets

Table 4.9 The RMC Ti–O coordination distribution percentages calculated for 20 configurations

Sample	4 coordinated (%)	5 coordinated (%)	6 coordinated (%)	Total
BTO	34.7 ± 0.7	51.4 ± 0.8	12.2 ± 0.4	4.74 ± 0.01
Nd BTO	30.1 ± 0.8	55.5 ± 1.1	12.4 ± 0.4	4.78 ± 0.01
Gd BTO	17.1 ± 0.2	65.1 ± 0.4	17.8 ± 0.4	5.01 ± 0.01
Er BTO	46.0 ± 0.7	42.1 ± 1.4	5.8 ± 0.7	4.47 ± 0.01
Yb BTO	44.6 ± 0.8	29.8 ± 1.1	3.7 ± 0.4	4.11 ± 0.01

In all samples there was some percentage of 3 coordinated Ti–O; in the case of Er BTO and Yb BTO these proportions were relatively large (5.9 and 19.8 % respectively)

function arises solely from Ti–O. This allows for the determination of the Ti–O coordination number (Table 4.8), which was calculated using three methods: the running coordination number method; the two Gaussian fit method (Fig 4.8); and by symmetrising the first peak about the maximum. The significantly smaller coordination number determined by reflecting the first peak about its maximum is an indication of the peak being skewed to larger distances.

By comparison, [35] determined Ti–O coordination numbers of 4.87 ± 0.15 and 5.05 ± 0.15 by fitting two Gaussians (at 1.91 and 2.13 Å) to their X-ray and neutron diffraction data respectively; they attributed the Gaussians to two different bond lengths in 5 coordinated square pyramid polyhedra seen in alkali titanium silicate glasses.

The Ti–O coordination number, and its distribution, was also determined from the RMC, as shown in Table 4.9. It is interesting to note that along with 4 and 5 fold coordinated Ti–O, the RMC suggests there is also a significant amount of 6 coordinated Ti–O. This is similar to the Ti–O coordination environment discovered [12] in $BaTiAl_2O_6$, albeit with a significantly higher ratio of 4 coordinated to 6 coordinated Ti.

The multiple coordination environments of Ti–O are also suggested by a bond angle analysis (Fig. 4.9). The O–Ti–O bond angle distribution is a very broad peak with a maximum at 90° (a bond angle that occurs in all 5 and 6 coordinated polyhedra) and a significant proportion at 109° (the tetrahedral bond angle). The O–O–O bond angle has both intrapolyhedral and interpolyhedral contributions; the peak at 60° is due to three O atoms in the same tetrahedra and/or due to three O atoms in the same 5 coordinated polyhedra; the peak at 90° could arise either due to oxygen atoms on either the same or adjacent 5 or 6 coordinated polyhedra; and

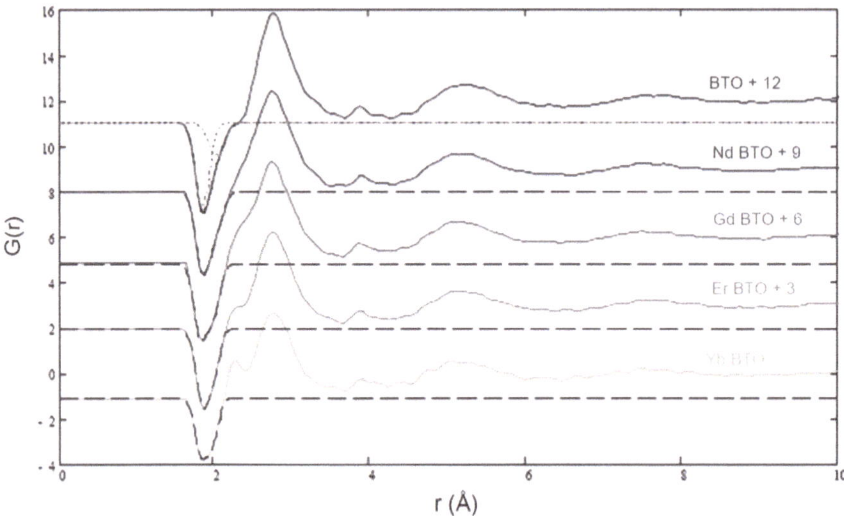

Fig. 4.8 The prefactors normalised $G(r)$'s for BTO glass and the rare earth glasses. The *black lines* are the double Gaussian fits to the first peak, with the two separate Gaussian's being shown (*dotted grey lines*) in the case of the pure BTO glass

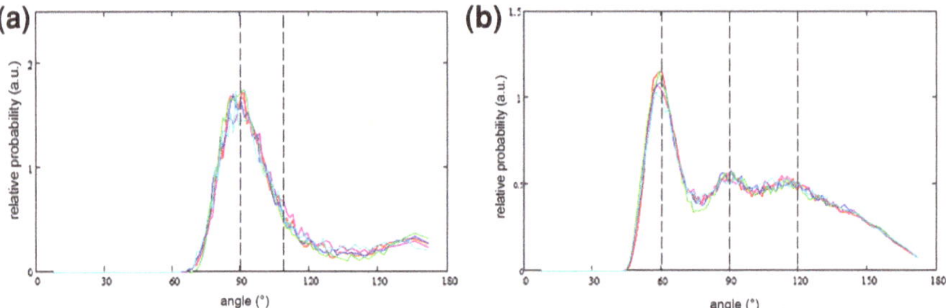

Fig. 4.9 RMC bond angle distributions for all samples for **a** O–Ti–O and **b** O–O–O. In both diagrams: BTO glass (*red*), Nd doped BTO (*blue*), Gd doped BTO (*green*), Er doped BTO (*pink*), and Yb doped BTO (*light blue*). The *broken black lines* are at **a** 90° and 109°, and **b** 60°, 90° and 120°

the peak at 120° can arise due to interpolyhedral contributions from either 4 or 5 coordinated Ti–O.

As mentioned in the results section, a thorough analysis of the XANES region is not possible. However a tentative comparison can be made with the results of Mountjoy et al. [22], who reported reference peak heights and positions shown in Table 4.10. The relative peak positions of the five samples (Table 4.7) imply 4 coordinated Ti is present. Although the XANES peak height is monochromator dependent, as Mountjoy et al. [22] also used an Si(111) monochromator a semi-quantitative comparison will be made. From Table 4.7 it can be seen that the relative peak heights of the five samples lie between the 4/5 and 6 coordinated heights

Table 4.10 The relative positions and heights of references samples reported by Mountjoy et al. [22]

Reference	Ti–O Coordination	Relative Position (eV)	Relative height
Ba_2TiO_4	4	3.50	0.60[a]
Na_2TiSiO_5	5	4.37[a]	0.55[a]
$ZrTiO_4$	6	4.76	0.15[a]
Rutile	6	5.40	0.24[a]
Anatase	6	5.70	0.20[a]

[a]Values that were not explicitly stated were graphically estimated

of Mountjoy et al. [22]. This implies that pure 4 coordination is improbable, and it seems possible the reduced peak height could indicate a degree of 6 coordination. This appears to support the conclusions of the RMC, although as highlighted the verification the XANES provides is of limited reliability due to the absence of reference measurements and a thorough energy calibration.

When considering the introduction of a dopant into a glass network, it is important to determine the influence of the dopant on the network. Initial observations of the relatively small variations in the neutron diffraction data and the invariability of the Ti K-edge real space peak position, suggest that the glass forming network does not substantially change; given the relatively low doping concentration (~3.5 %) this is not unexpected. Assuming that the first peak is still due purely to Ti–O correlations, the Ti–O coordination number can be calculated (Table 4.8). From this it appears that the Ti–O coordination number is significantly higher than pure BTO for the larger rare earth ions (Nd and Gd), and significantly lower for the smaller rare earth ions (Er and Yb). Due to the disagreement between the Gd doped $F(q)$ and the back Fourier transform of the cutoff $G(r)$, it is probable that the Ti–O coordination determined is higher than the real coordination. In the case of Er and Yb doped glasses however, it is probable that the apparently decreased Ti–O coordination occurs due to a different reason.

In order to highlight the subtle differences between the total structure factors and total correlation functions, Fig. 4.10 shows the effects of subtracting the doped glass structure factors from the pure BTO and Fourier transforming. From this it becomes evident that the small increase in the doped total correlation functions at ~2.3 Å moves to lower distances with decreasing ion size, from ~2.42 Å for the Nd doped to ~2.27 Å for the Yb doped glass. When considering the RMC partial pair correlation functions (Fig. 4.11) there appear to be three different correlation functions that could contribute to the total correlation function at this distance; the Ti–O, the Ba–O, and the RE–O correlations. As this feature varies with ion size it seems probable that it arises due to the RE–O correlation; this is implied by the RMC (Fig. 4.11). Although the RMC also suggests that the Ba–O partial pair correlation function also sees an increase in the low r correlations with decreasing RE ion size, it is probable that this is an artefact due to the small number of constraints; it would be interesting to attempt to determine if this is the case by further constraining the RMC with X-ray diffraction data.

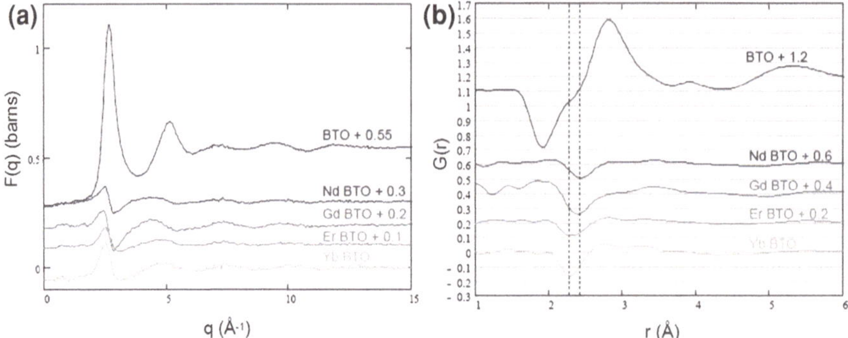

Fig. 4.10 **a** The pure BTO total structure minus the structure factors of the different doped glasses, Nd (*blue*), Gd (*green*), Er (*pink*) and Yb (*light blue*), shown with the pure BTO structure factor (*red*) for comparison. **b** The Lorch modified $G(r)$'s determined by Fourier transforming the structure factors from **a**. The *dotted black lines* are at $r = 2.27$ and 2.42 Å

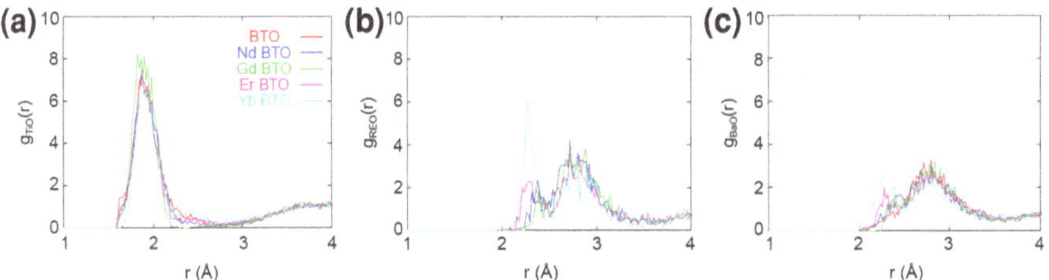

Fig. 4.11 The partial pair correlation functions for BTO (*red*), Nd (*blue*), Gd (*green*), Er (*pink*) and Yb (*light blue*), for **a** Ti–O, **b** RE–O, and **c** Ba–O determined by RMC. There is a bifurcation of the RE–O peak which is most significant for the smallest ionic radii (Yb). This bifurcation probably occurs due to the limitations of the underdetermined RMC; what it implies is a shortening RE–O bond length with decreasing RE ion size

The change in RE–O bond length is also evident from the RE L_3-edge EXAFS data (Fig. 4.5); the peak position changes by ~0.12 Å from the Nd doped glass to the Yb doped glass. For comparison, the effective ionic radii [25] of Nd and Yb are 1.109 and 0.985 Å. This change in the RE–O bond length explains the apparent reduction of the Ti–O coordination determined from the neutron diffraction data, as for the smaller rare earth ions the assumption that only that Ti–O coordination shell is isolated breaks down.

As discussed in Sect. 4.1 the clustering of rare earth ions strongly influences their fluorescence properties. Therefore the number of RE–O–RE was calculated, based upon the first minima in the RE–O partial correlation function. The results (Table 4.11) suggest that at doping percentage of 3.5 %, only ~10 % of rare earth atoms demonstrate clustering. Therefore this property, combined with the low phonon energy [17], suggests that BTO glass is a good candidate for fluorescent RE

Table 4.11 An indication of the clustering of rare earth atoms shown by the percentage of RE atoms in RE–O–RE bonds

Sample	RE–O coordination number	Percentage of RE atoms in RE–O–RE bonds (%)
Nd BTO	9.23 ± 0.03	10.1 ± 0.2
Gd BTO	8.94 ± 0.05	9.9 ± 0.2
Er BTO	8.63 ± 0.05	9.5 ± 0.2
Yb BTO	9.08 ± 0.08	10.2 ± 0.2

doping. An interesting further study would be to determine the effect on the rare earth clustering of increasing the doping percentage, which may be achieved by the substitution of barium by calcium [21].

4.5 Conclusions

To summarise the key results from this combined neutron diffraction, XAS, and RMC study on BTO and 3.5 % rare earth doped BTO glass are: 4, 5 and to a lesser extent 6 coordination Ti–O polyhedra form the underlying glass network, similar to the coordination of Ti–O present in other glasses [12]; the rare earth oxygen bond length decreases with decreasing rare earth ionic radius; and only 10 % of the rare earth atoms are found in RE–O–RE bonding.

References

1. Akishige Y, Fukano K, Shigematsu H (2003) New ferroelectric $BaTi_2O_5$. Jpn J Appl Phys 42(2–8):L946–L948
2. Akishige Y, Fukano K, Shigematsu H (2004) Crystal growth and dielectric properties of new ferroelectric Barium Titanate: $BaTi_2O_5$. J Electroceram 13(1–3):561–565
3. Arai K, Namikawa H, Kumata K, Honda T, Ishii Y, Handa T (1986) Aluminum or phosphorus co-doping effects on the fluorescence and structural properties of neodymium-doped silica glass. J Appl Phys 59(10):34303436
4. Borrelli NF, Layton MM (1971) Dielectric and optical properties of transparent ferrorelectric glass–ceramic systems. J Non-Cryst Solids 6(3):197–212
5. Bush TS, Gale JD, Catlow RA, Battle PD (1994) Self-consistent interatomic potentials for the simulation of binary and ternary oxides. J Mater Chem 4(6):831–837
6. Dent AJ, Cibin G, Ramos S, Smith AD, Scott SM, Varandas L, Pearson MR, Krumpa NA, Jones CP, Robbins PE (2009) B18: a core XAS spectroscopy beamline for Diamond. J Phys Conf Ser 190:012039
7. Dianoux A-J, Lander G (eds) (2003) Neutron data booklet, 2nd edn. Old City Publishing, Philadelphia
8. Farges F, Brown GE, Rehr JJ (1996) Coordination chemistry of Ti(IV) in silicate glasses and melts: I. XAFS study of titanium coordination in oxide model compounds. Geochim Cosmochim Acta 60(16):3023–3038
9. Farges F, Brown G, Rehr J (1997) Ti K-edge XANES studies of Ti coordination and disorder in oxide compounds: comparison between theory and experiment. Phys Rev B 56(4):1809–1819

10. Fischer HE, Cuello GJ, Palleau P, Feltin D, Barnes AC, Badyal YS, Simonson JM (2002) D4c: a very high precision diffractometer for disordered materials. Appl Phys A 74(1):160–162
11. Huggins ML, Mayer JE (1933) Interatomic distances in crystals of the alkali halides. J Chem Phys 1(9):643
12. Kidkhunthod P (2012) The structure of Rare-Earth gallate and aluminate glasses determined by neutron and X-ray diffraction and spectroscopy. Ph.D. thesis, University of Bristol
13. Lewis GV, Catlow CRA (1985) Potential models for ionic oxides. J Phys C: Solid State Phys 18(6):1149–1161
14. Lines ME (1977) Microscopic model for a ferroelectric glass. Phys Rev B 15(1):388–395
15. Mason WP, Matthias BT (1948) Theoretical model for explaining the ferroelectric effect in Barium Titanate. Phys Rev 74(11):1622–1636
16. Masuno A, Inoue H, Yu J, Arai Y, Atsubo F (2008) Thermal stability and optical properties of Er^{3+} doped $BaTi_2O_5$ glasses. Adv Mater Res 39–40:243–246
17. Masuno A, Inoue H, Yu J, Arai Y (2010) Refractive index dispersion, optical transmittance, and Raman scattering of $BaTi_2O_5$ glass. J Appl Phys 108(6):063520
18. McCauley D, Newnham RE, Randall CA (1998) Intrinsic size effects in a Barium Titanate glass–ceramic. J Am Ceram Soc 81(4):979–987
19. McGreevy RL, Pusztai L (1988) Reverse Monte Carlo simulation: a new technique for the determination of disordered structures. Mol Simul 1(6):359–367
20. Meyerhofer D (1958) Transition to the ferroelectric state in Barium Titanate. Phys Rev 112(2):413–423
21. Moriyoshi C, Kuroiwa Y, Masuno A, Inoue H (2012) Site-selective calcium substitution in $BaTi_2O_5$: effect on the crystal structure and the ferroelectric phase transition. J Phys Soc Jpn 81(1):014706
22. Mountjoy G, Pickup DM, Wallidge GW, Anderson R, Cole JM, Newport RJ, Smith ME (1999) XANES study of Ti coordination in heat-treated $(TiO_2)_x$ $(SiO_2)_{1\text{-}x}$ xerogels. Chem Mater 11(5):1253–1258
23. Mughabhab SF (2006) Altas of neutron resonances, 5th edn. Elsevier, Amsterdam
24. Ravel B, Newville M (2005) ATHENA, ARTEMIS, HEPHAESTUS: data analysis for X-ray absorption spectroscopy using IFEFFIT. J Synchrotron Radiat 12(Pt 4):537–541
25. Shannon R (1976) Revised effective ionic radii and systematic studies of interatomic distances in halides and chalcogenides. Acta Crystallogr A A32(SEP1):751–767
26. Skinner LB, Barnes AC, Crichton W (2006) Novel behaviour and structure of new glasses of the type Ba–Al–O and Ba–Al–Ti–O produced by aerodynamic levitation and laser heating. J Phys Condens Matter 18(32):407–414
27. Skinner LB, Barnes AC, Salmon PS, Fischer HE, Drewitt JW, Honkimäki V (2012) Structure and triclustering in Ba–Al–O glass. Phys Rev B 85(6):064201
28. Smith W, Forester T (1996) DL POLY 2.0: a general-purpose parallel molecular dynamics simulation package. J Mol Graph 14(3):136–141
29. Sun KH (1947) Fundamental condition of glass formation. J Am Ceram Soc 30(9):277–281
30. Thompson A, Attwood D, Gullikson E, Howells M, Kim K-J, Kirz J, Kortright J, Lindau I, Liu Y, Pianetta P, Robinson A, Scofield J, Underwood J, Williams G, Winick H (2009) X-ray data booklet, 3rd edn. Lawrence Berkeley National Laboratory, University of California, Berkeley
31. Weber R, Hampton S, Nordine PC, Key T, Scheunemann R (2005) Er^{3+} fluorescence in Rare-Earth aluminate glass. J Appl Phys 98(4):043521
32. Yamane M, Asahara Y (2000) Glasses for photonics. Cambridge University Press, Cambridge
33. Yao K, Zhang LY, Yao X, Zhu WG (1997) Preparation and properties of barium titanate glass–ceramics sintered from sol–gel-derived powders. J Mater Sci 32(14):3659–3665
34. Yu J, Arai Y, Masaki T, Ishikawa T, Yoda S, Kohara S, Taniguchi H, Itoh M, Kuroiwa Y (2006) Fabrication of $BaTi_2O_5$ glass–ceramics with unusual dielectric properties during crystallization. Chem Mater 18(8):2169–2173
35. Yu J, Kohara S, Itoh K, Nozawa S, Miyoshi S, Arai Y, Masuno A, Taniguchi H, Itoh M, Takata M, Fukunaga T, Koshihara S, Kuroiwa Y, Yoda S (2009) Comprehensive structural study of glassy and metastable crystalline $BaTi_2O_5$. Chem Mater 21(2):259–263

Chapter 5
Liquid–Liquid Transitions in Y_2O_3–Al_2O_3 System

5.1 Introduction

As discussed in Chap. 2, there is significant controversy about the observed liquid–liquid phase transition in the yttria aluminates. To briefly recap, Aasland and McMillan [1] observed clear spherical inclusions in glassy samples in a composition range of $0.24 < x < 0.32$, which they attributed to a second glass phase. From this they inferred the occurrence of a liquid–liquid phase transition which was interrupted during quenching. The glassy nature of these inclusions has since been reported as crystalline [3, 8, 11]. In this thesis, as is common in the literature, yttria aluminate compositions $(Y_2O_3)_x(Al_2O_3)_{1-x}$ will be denoted AY(x), with the exception of $x = 0.375$ which will be denoted as YAG.

Following this was the observation by Greaves et al. [5] of an in situ liquid–liquid phase transition between a high density liquid using aerodynamic levitation combined with SAXS/WAXS. They reported that AY20 transitioned from a high density liquid (HDL) above 1,788 K to a low density liquid (LDL) below it. It is important to note that Greaves et al. [5] did not observe any hint of a liquid–liquid phase transition occurring for any other AY composition, and that this composition is not within the solid polyamorphism range of Aasland and McMillan [1]; however it should be mentioned that Aasland and McMillan did observe nucleation and growth of an apparent second liquid phase at AY20 before crystallisation intervened.

The evidence for this apparently first order transition between liquid states with a 4 % density difference comes from three separate indicators.

In the WAXS they saw a small, apparently discontinuous, increase in the $S(q)$ first peak position and intensity at ~1,788 K. They attributed this to a shift from edge shared tetrahedra in the HDL to corner shared tetrahedra in the LDL. The magnitude of this shift is ~0.05 Å^{-1}. However in the high energy X-ray diffraction experiment of Barnes et al. [3], which again used in situ aerodynamic levitation and studied an even greater temperature range at smaller temperature separations, this discontinuous change in peak intensity and position was absent. Some of the other results contained within Barnes et al. [3] are discussed in Sects. 5.3 and 5.4.

T. Farmer, *Structural Studies of Liquids and Glasses Using Aerodynamic Levitation*, Springer Theses, DOI: 10.1007/978-3-319-06575-5_5,

© Springer International Publishing Switzerland 2015

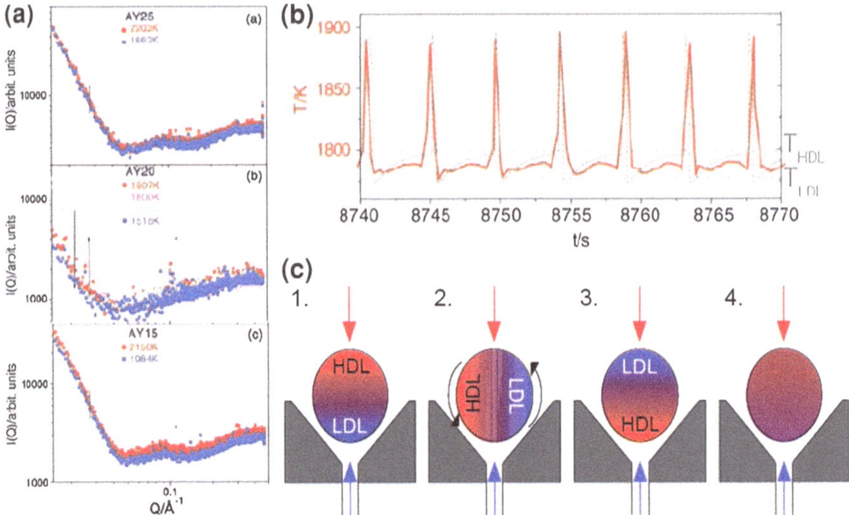

Fig. 5.1 **a** SAXS data of Greaves et al. [5] which suggests a significant increase in sample inhomogeneity for AY20 at temperatures of 1,786 and 1,800 K, straddling the T_{LL} of 1,788 K. **b** Pyrometric spikes of Greaves et al. [5] attributed to a 'polyamorphic rotor'. **c** A model describing the behaviour of a 'polyamorphic rotor', with the laser heating and the cold gas depicted with *red* and *blue arrows* respectively. As the temperature of the sample is reduced from above T_{LL}, the *bottom* of the sample, which is ~50 K colder than the *top*, starts to transform into the low temperature, low density liquid. This causes a mechanical instability which results in the sample rotating so that the LDL is situated at the *top* of the sample. The laser heating then causes the LDL to heat and form HDL, while the opposite occurs to the bottom of the sample, resetting the process. From Greaves et al. [5]. Reprinted with permission from AAAS

The SAXS of Greaves et al. [5] detected a significant increase (>100 % at the smallest q) in the scattering intensity below $q = 0.05$ Å^{-1} over a small temperature range (~1,760–1,810 K). This is indicative of a significant increase in inhomogeneity of the scattering length density, which was attributed to the formation of the LDL centred on a temperature of $T_{\mathrm{LL}} = 1{,}788$ K. This is shown in Fig. 5.1a.

The final suggestion of a liquid–liquid phase transition by Greaves et al. [5] came from atypical pyrometric curves. Generally upon completely melting a sample to a liquid the pyrometry becomes relatively smooth and constant, with temperature oscillations only of the order of a few Greaves et al. [5], however, witnessed large, repeated temperature spikes (~170 K) with a period of ~4 s (Fig. 5.1b). They attributed this to a mechanism they dubbed as a 'polyamorphic rotor', which is best described diagrammatically (Fig. 5.1c). As their experimental setup (like that used for this body of work-see Sect. 3.1) consisted of a single laser incident on the top of the sample, there is a temperature gradient of ~50 K from the top to the bottom. This can lead to the top of the sample being at $T > T_{\mathrm{LL}}$, and so consists of HDL, while the bottom of the sample is at $T < T_{\mathrm{LL}}$, and so consists of LDL. This results in an instability which causes the sample to flip, resulting in a large reduction in measured temperature as the cold LDL becomes incident under the pyrometer (also situated at the top of the sample). Thermodynamic considerations based on the 'polyamorphic rotor' observations resulted in Greaves et al. [5] determining an enthalpy change of $\Delta H_{LL} = 34 \pm 8$ kJ mol^{-1}; however Barnes et al. [3] point out that an enthalpy of this magnitude would cause an observable plateau in the pyrometry of liquid AY20 fast quenched from $T > T_{LL}$, which is not apparent.

In order to resolve the issues surrounding the liquid–liquid phase transition in the yttria aluminates, it is essential that these suggested indications are thoroughly corroborated. If a liquid–liquid phase transition can be definitively confirmed in the yttria aluminates, it would be the first conclusive demonstration of such phenomena. This would in turn further stimulate research into other possible systems exhibiting polyamorphous behaviour.

5.2 Methods

Samples of AY20, AY25, AY30, and YAG were formed from Al_2O_3 (99.99 % pure) and Y_2O_3 (99.99 % pure) using the two ball method described in Sect. 3.1. In order to reduce the effect of oxygen loss from the samples, all levitation was performed in an oxygen atmosphere. After rapidly quenched the samples from the liquid, the AY20 was completely crystalline, the AY25 (which is on the edge of the glass forming region) formed a turbid glass, and the AY30 formed a glass. The sample diameters were all approximately 2.9 mm, which equates to sample masses of ~45 mg.

SANS was carried out on the D22 diffractometer at the ILL. Samples were measured at a neutron wavelength of 4.5 Å and a collimator-sample and sample-detector distance of 8 m giving 0.008 Å^{-1} $\leq q \leq$ 0.185 Å. Each temperature was measured for 1,000 s divided into 100 s scans in order to check stability. An oxygen/argon mixture (3 %/97 %) was used as a levitation gas, again in order to reduce oxygen loss during levitation. B_4C and Cd masks placed up-beam were used to eliminate scattering from the vanadium nozzle (Fig. 5.2). Diffraction measurements were also taken of the empty chamber and the chamber with gas flow for correction purposes.

As AY20 was the composition for which a liquid–liquid phase transition had previously been observed by Greaves et al. [5] this composition had SANS measurements taken at a large number of temperatures. The temperatures for each sample are shown in Table 5.1. All samples were measured in the solid for comparison; for YAG and AY20 the room temperature form was crystalline, for AY30 it was glassy, and AY25 was a turbid glass probably containing sub micron sized crystals [11]. The temperatures listed were determined by a single wavelength (850 nm) pyrometer and have not been corrected for emissivity. The emissivity of the yttria aluminates is approximately 0.92 [10], which corresponds to a difference between the measured and actual temperatures of ~30 K at 2,173 K. The purported liquid–liquid phase transition [5] occurs in a range of ~50 K centred on 1,788 K. This range is well covered by the run temperatures, even considering the small uncertainty due to not having an exact value for the emissivity.

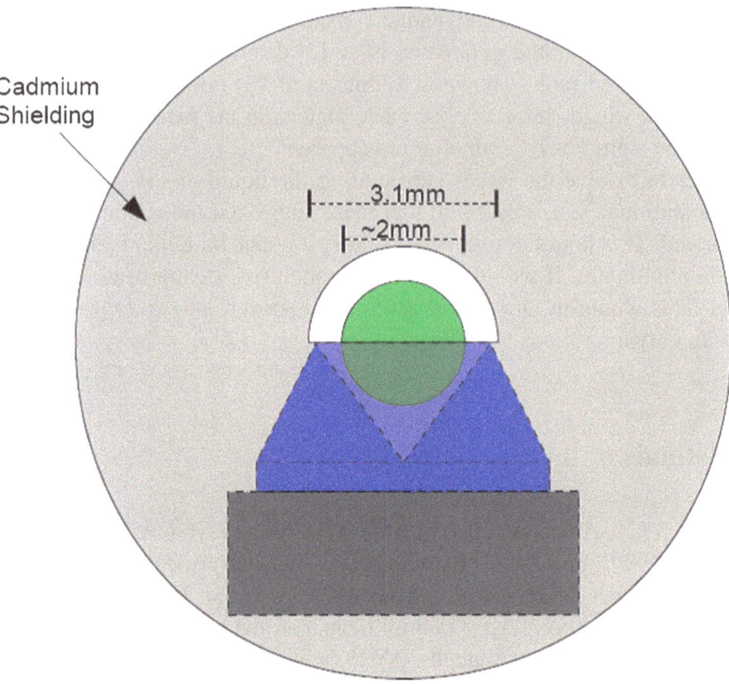

Fig. 5.2 A schematic of the sample setup shown from the direction of the beam. The beam overlaps the semicircular hole but is not shown for clarity

Table 5.1 Sample temperatures (uncorrected for emissivity) at which SANS measurements were taken

Sample	Temperatures (K)
AY20	2023, 1973, 1923, 1883, 1853, every 10 K from 1833 to 1743, every 20 K from 1743 to 1583, room temperature
AY25	2073, 1973, 1873, 1773, 1673, 1623, 1573, 1523, room temperature
AY30	Room temperature
YAG	2173, 2073, 1973, 1873, 1773, 1673, room temperature

The analysis of the SANS was performed using the GRASP software package [4] which is designed for use on data collected from D22. The empty beam transmission measurement was used in order to determine the beam centre, which is obviously required to calculate the radial distribution function from the 2D intensity distribution. A mask was also applied to the sample data in order to remove the effects of the direct beam. As can be seen in Fig. 5.3 the effect of varying the size of this mask does not significantly affect the sample data above 0.01 Å^{-1}.

An important consideration in the data analysis is the amount of beam transmitted. In this particular case, as with most if not all levitation experiments, the neutron beam completely covers the section of the sample above the nozzle. The setup

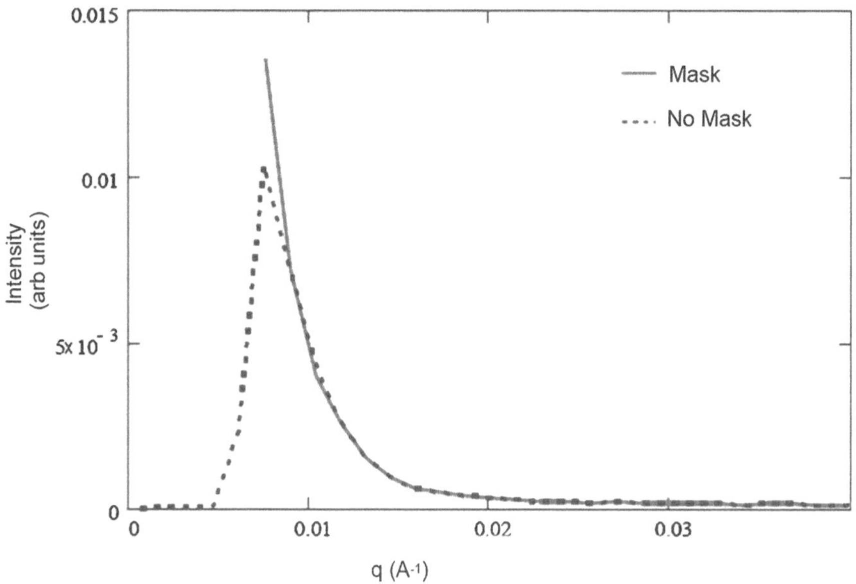

Fig. 5.3 A graph of I versus q comparing the effects of masking the data for AY20 at 1,783 K. The effect of masking the data is insignificant for $q > 0.01$ Å^{-1}

specific to this experiment is shown in Fig. 5.2. The amount of sample that protrudes above the nozzle is obviously dependent on a number of factors, including the flow rate, sample mass, sample volume, and nozzle shape, although this usually equates to approximately 50 % of the total sample volume. Due to the spherical sample shape the transmission will obviously vary across the height of the sample, however the average transmission can be calculated to be 96 %, assuming that a hemisphere is illuminated. Over the duration of levitation and laser heating there was a small change in the transmission due to sample mass loss, which was approximately 10 %. This was confirmed by comparing the total count rates at the start and end of a sample run.

As well as SANS measurements, a separate systematic study of the oscillatory behaviour of AY20 and AY30 was performed in order to test for the occurrence of a polyamorphic rotor. These measurements were taken on the aerodynamic levitator setup at the University of Bristol. Sample temperatures were again measured using a single wave pyrometer and videos were taken from two angles (Fig. 5.4).

5.3 Results

Figure 5.5 shows a log–log plot of the SANS intensity against q for AY20 at a number of temperatures. It is apparent that there is no significant difference in the intensity of the scattering at any temperature. As is expected if the small angle

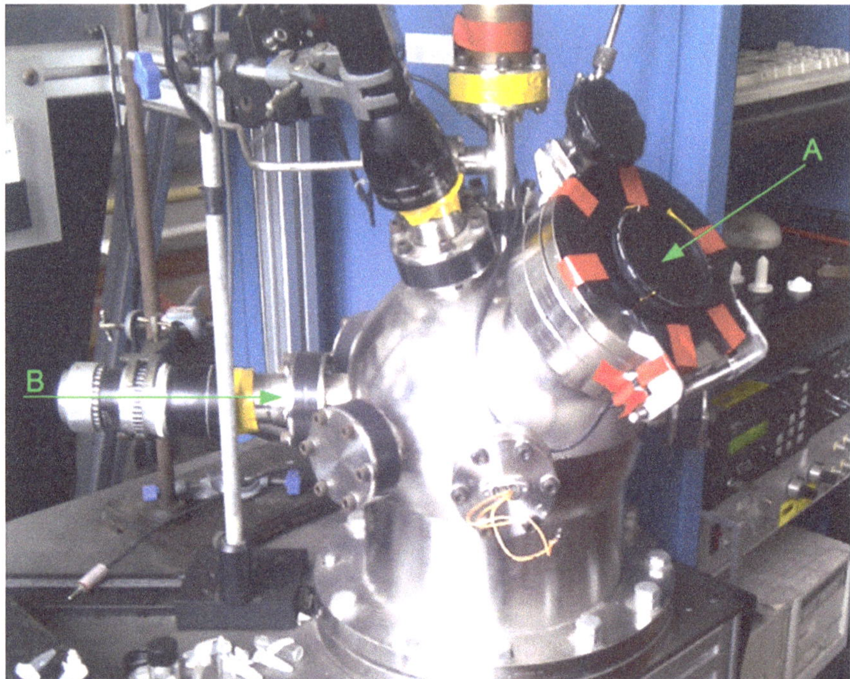

Fig. 5.4 A picture of the levitator chamber with the two camera positions indicated

scattering arises purely from thermally induced density fluctuations, there is a slight increase in the intensity with temperature. This can also be seen in the AY25 and YAG data (Fig. 5.5). The only sample which demonstrate a slightly different intensity rise are those of the AY25 cloudy glass, which is shown in Table 5.2. This small intensity rise is related to the turbidity of this glass, which occurs due to microscopic inhomogeneity.

Another notable feature when comparing to the SAXS data is that there is only a small difference between the intensities of the three compositions, as shown in the insert in Fig. 5.5. This is in stark contrast to Greaves et al. [5] where the small angle scattering intensity in AY25 at all temperatures appears to be up to ten times larger than that of the AY20 at temperatures away from the purported liquid–liquid transition. Considering that at these temperatures the increase in small angle scattering is solely due to thermally induced density fluctuations it is surprising that these would increase significantly with a small change in composition. The Greaves et al. [5] SAXS intensity for all AY25 temperatures is in fact even greater than in AY20 at the transition temperature. Even assuming the SAXS is unnormalized with respect to the sample volume, the AY25 sample diameter would have to be ~2.15 times greater to account for the difference. This anomaly in the SAXS intensity between samples could be indicative of the difficulties of SAXS on levitated samples which will be discussed later.

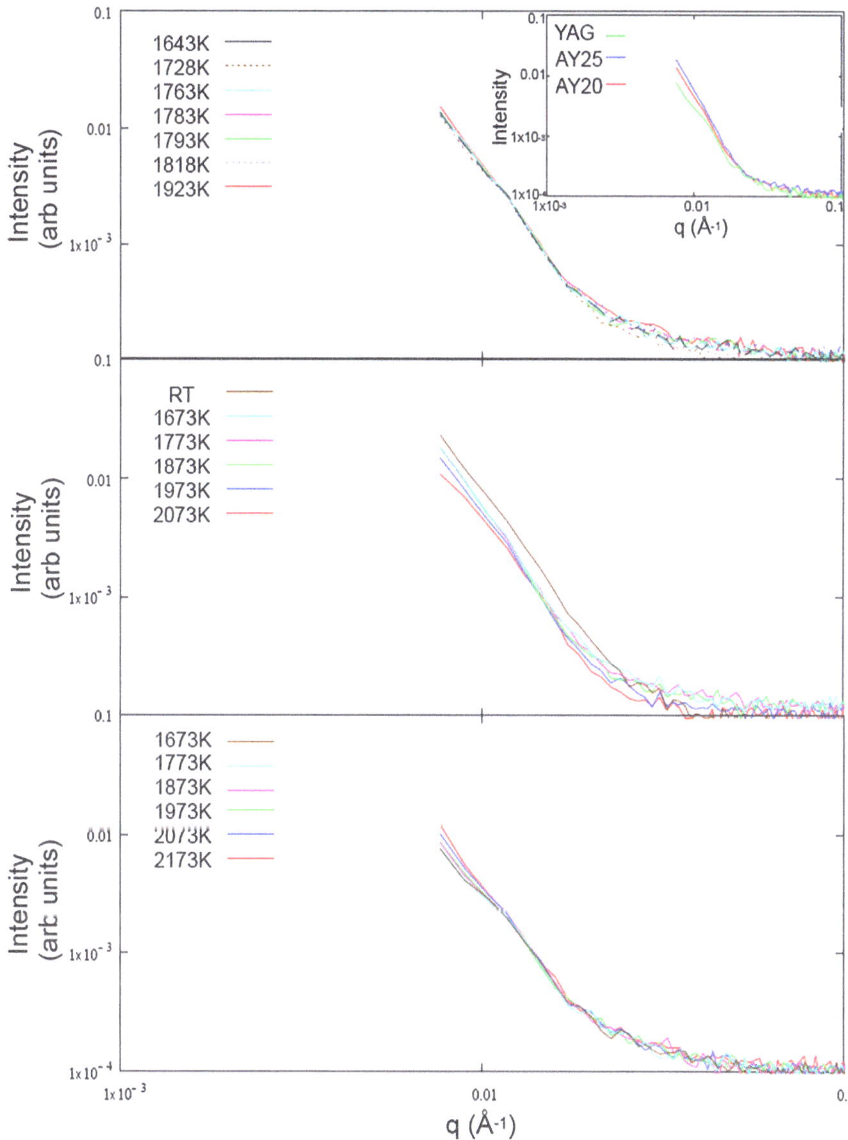

Fig. 5.5 Log–Log plots of the SANS intensity for AY20 (*Top*), AY25 (*Middle*), and YAG (*Bottom*) at various temperatures representative of all measured temperatures. *Insert* comparing the SANS intensity for the three compositions at $T = 1{,}783$ K for AY20 and $T = 1{,}773$ K for AY25 and YAG. The error on the intensity is <1 % in all cases. There is no significant increase in SANS intensity at any temperature for any composition

The integrated SANS intensities for AY20, AY25, and YAG are compared with the SAXS data from Greaves et al. [5] in Fig. 5.6. The significant increase in integrated SAXS centred around 1,788 K is not apparent in the SANS data at any

Table 5.2 Power law gradients determined for the 0.01–0.03 Å^{-1} region of four different runs

Sample	Temperature (K)	Gradient
AY20	1,643	−3.63
AY25	1,673	−3.74
AY25	300 (RT)	−4.03
AY30	300 (RT)	−3.57

The smallest R^2 value for the four gradients was 0.98. The only significant difference for all samples and temperatures is for the AY25 cloudy glass sample

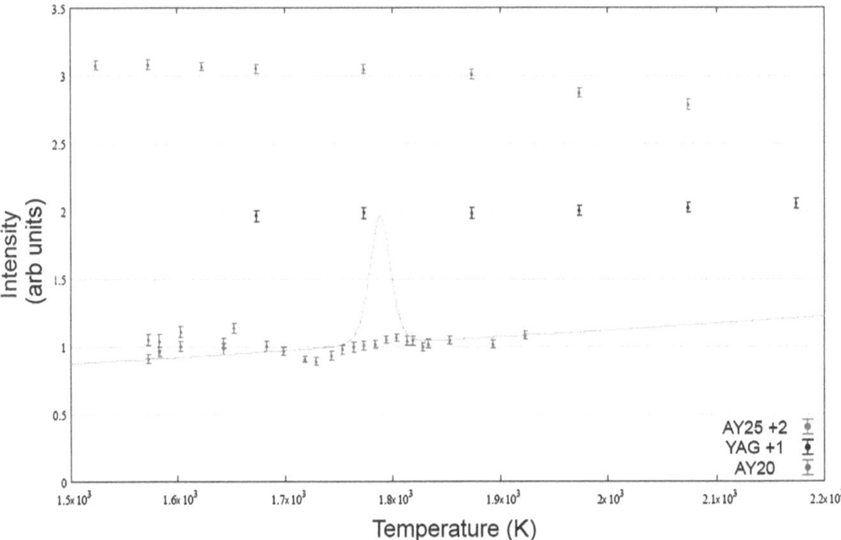

Fig. 5.6 The integrated SANS intensity for AY20, AY25 and YAG between $q = 0.03$ Å^{-1} and $q = 0.05$ Å^{-1} (the same bounds as Greaves et al. [5]) normalised to the average for each composition. No significant variation of intensity was observed by altering the integration bounds up to 0.18 Å^{-1}. The *green line* shows the double sigmoid fit to the data of Greaves et al. As Greaves et al. have used arbitrary units, the *peak* shown was determined based upon scaling relative to the average intensity away from the *peak*. While the AY20 does not closely follow the expected increase of integrated intensity with temperature, it is apparent that there is no significant increase in integrated intensity at any temperature

temperature for any composition. The AY25 data shows a decrease in the integrated intensity as the temperature increases. It is likely that this occurs due to the fact that the AY25 sample was observed to become particularly oblate at high temperatures.

5.4 Discussion

As previously mentioned there is an inherent uncertainty in the transmission which is also time dependent because of the sample mass loss. In SANS from levitated samples, this uncertainty is relatively small and the time dependence is

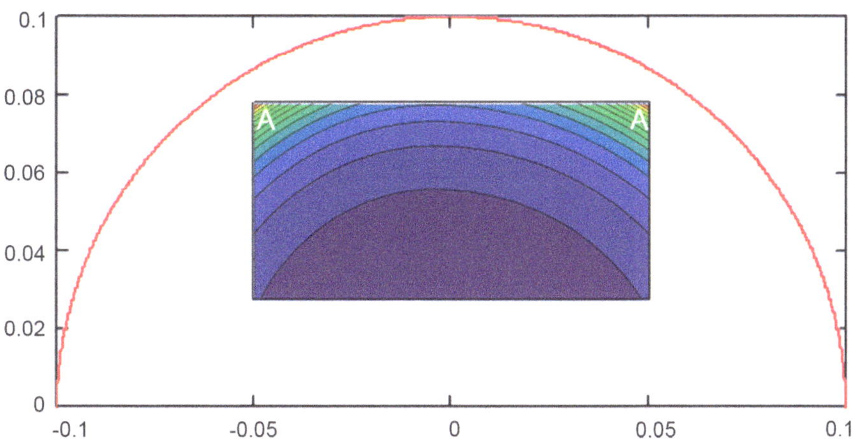

Fig. 5.7 X-ray transmission contours for a 0.5 mm by 1 mm X-ray beam incident on a 2 mm sample spherical sample required for an average transmission of 3 %, as was reported by Greaves et al. [5]. The contours represent a transmission of 1 % in the largest *purple region*, increasing in 1 % increments to 16 % at the smallest *orange region*. The two areas of the majority of transmission are indicated by the letter A

weak. For example a reduction in the sample diameter from 2 to 1.6 mm, which is considerably greater than is expected, would lead to a change in the transmission from 96 to 97 % for AY20. This is significantly different to the transmission behaviour in SAXS due to two factors; the transmission is significantly smaller due to the interaction, and the X-ray beam only illuminates a small volume of the upper hemisphere of the sample. This means that in considering the uncertainty in beam position due to both the positioning precision and sample height changes there is a significant error on the transmission for the SAXS. Figure 5.7 shows the X ray transmission for a spherical AY20 sample of 2 mm diameter with a beam of 0.5 mm height and 1 mm width. The position of the beam is such that the average transmission is 3 %, which is the reported transmission of Greaves et al. [5]. From Fig. 5.7 it is obvious that a significant portion of the transmission occurs from two relatively small regions of the beam, labelled A. This is reflected in the changes in transmission shown in Table 5.3. Although using the error in sample instability of Greaves et al. [5] results in a relatively small error in the transmission, other reported values [2, 9] of sample instability from high precision setups suggest that Greaves et al. [5] may have underestimated this key factor. When considering the combined effects of sample mass loss over time, changes in surface tension leading to varying sphericity, and sample instability inherent in levitation, there is a very significant uncertainty in the SAXS transmission. Ultimately this results in a considerable uncertainty in the integrated SAXS intensity.

However a complication arises in the case of SANS due to the beam overlapping the sample. It is possible that this contribution to the scattering, which of course does not occur in SAXS, could swamp any increase in scattering due to the nucleation of a second liquid phase. The levitated sample can effectively be

Table 5.3 The error on the transmission due to sample instability calculated for SAXS from a 2 mm levitated AY20 sample, with a beam of 1 mm width and 0.5 mm height positioned such that the average transmission is 3 %

Sample instability			
Vertical (μm)		Horizontal (μm)	Percentage error in the transmission (%)
A	±2.3		2
B	±50	±0.9	49
		±50	
C	±25	±25	22

The sample stability in row A is that reported by Greaves et al. [5], although this is significantly lower than those reported for aerodynamic levitation [2] and combined aerodynamic and electrostatic [9] levitation shown in row B. Row C is an intermediate sample stability which still causes a significant error in the transmission

considered as a single sphere of radius ≈2 mm with 100 % contrast. By contrast any nucleation of a second liquid phase would be a large collection of microscopic spheres with a small contrast [5]. In the case of the $(Y_2O_3)_x(Al_2O_3)_{1-x}$ system the density difference between the two liquid phases has previously been reported [1] to be 4 %, and so this value was used for the contrast.

In the case of a monodisperse dilute collection of spheres, the spherical symmetry of the contrast factor leads to a simple form for the scattered intensity [6, 7]:

$$I(q) = N(\Delta\rho)^2 V^2 \left[\frac{3(\sin qR - qR\cos qR)}{(qR)^3}\right]^2 \tag{5.1}$$

where N is the number of particles, $\Delta\rho$ is the contrast factor, V is the volume of the scattering particle, and R is the particle radius. This function has sharp troughs for $qR = 4.493, 7.725, \ldots$ as shown in Fig. 5.8. In the case of the single large sphere these troughs are sufficiently sharp and close together that standard instrumental resolution will lead to a line with q^{-4} dependence (Fig. 5.8). For a collection of a monodisperse dilute small spheres the oscillations in the intensity would be visible in a SANS measurement. However it is obvious that the nucleation sites of a second liquid phase would be significantly polydisperse. This requires a simple modification [6] of (1) to include a size distribution, D(R):

$$I(q) = \int_0^\infty D(R)N(\Delta\rho)^2 V^2 \left[\frac{3(\sin qR - qR\cos qR)}{(qR)^3}\right]^2 dR \tag{5.2}$$

The introduction of a size distribution has the effect of smoothing out the oscillations, as shown in Fig. 5.8 where a Gaussian distribution has been assumed. Using these two simple models it is possible to calculate the ratio of L–L inclusion scattering to sample scattering for a variety of different packing fractions, φ, and average inclusion radii, $\langle R\rangle$ (Table 5.4). These calculations suggest that a rise due to a liquid–liquid phase transition would be visible if the following is satisfied:

$$\frac{\varphi}{\langle R\rangle} \geq 2\times 10^{-5}\ \text{Å}^{-1}$$

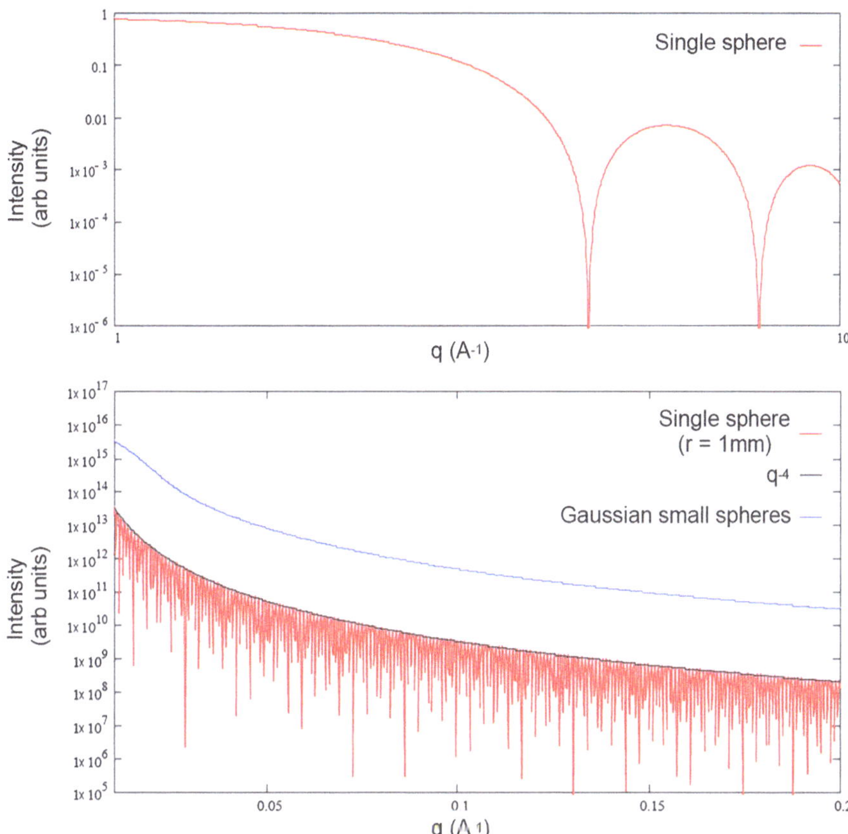

Fig. 5.8 *Top* theoretical SANS intensity from a single sphere calculated from (1). *Bottom* theoretical SANS intensity from a single large sphere and for a Gaussian distributed collection of small spheres. The small spheres line has been moved up for clarity

Table 5.4 Examples of the ratio of L–L inclusion scattering to sample surface scattering for different inclusions radii and packing fractions

Average radius of inclusions (Å)	Packing fraction	Intensity ratio
200	0.01	0.8
	0.05	4
	0.2	16
	0.5	40
100	0.01	1.6
	0.5	80
500	0.01	0.32
	0.5	16

All calculations assume a dilute system even though it is not valid for large packing fractions, as including the interparticle scattering will only cause an increase in the scattering at a q specific to the interparticle distance. For purposes of comparison the average radius of the included amorphous phase of Aasland and McMillan [1] was ~5 μm

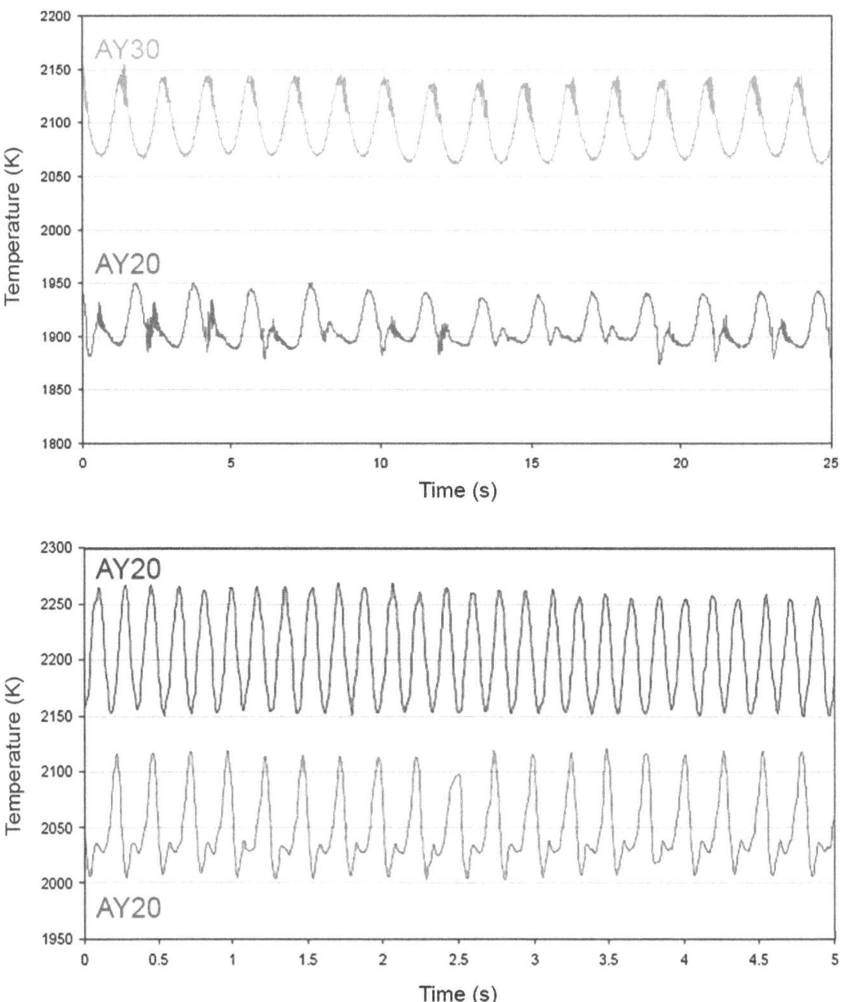

Fig. 5.9 Pyrometric traces of a similar nature to that representing a 'polyamorphic rotor' [5]. All traces are from AY20 samples except the *green trace* which is from an AY30 sample. The AY20 trace in the *top graph* was recorded during the video discussed in Fig. 5.10. This behaviour has been observed at a wide range of temperatures for a variety of different levitated liquids

which corresponds to a 32 % increase in the scattering intensity with the occurrence of an L–L phase transition. An increase of this magnitude would be significant enough that it would be visible on the integrated SANS intensity plot shown in Fig. 5.6. By contrast the largest deviation for the AY20 data from the linear thermal background behaviour expected is <20 %, and in the Greaves et al. [5] transition range of ~1,750–1,820 there are no deviations >5 %.

Figure 5.9 shows some pyrometric traces of both AY20 and AY30 at several temperatures. These behaviours are reminiscent of that reported by Greaves et al. [5] as indicative of a 'polyamorphic rotor'. They exhibit large repeated variations

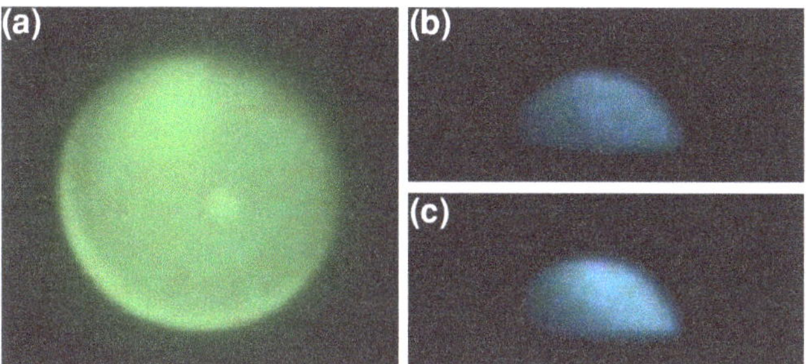

Fig. 5.10 **a** A still from camera angle *A* in Fig. 5.4 during pyrometry shown in Fig. 5.9. A *small bright spot* is clearly visible near the centre of the sample. The *bright patch* near the *top* of the sphere is the area on which the laser is incident. **b**, **c** Two stills 0.95 s apart taken from camera angle *B* in Fig. 5.4 during the same period as (**a**). While a general increase in brightness can be seen in **c** relative to **b** due to an increase in temperature of the side visible to the camera, it is not possible to see the *small bright spot*; therefore in certain circumstances the source of the abnormal pyrometric behaviour might not be visible. The sample diameter is ~2.5 mm, making the diameter of the *bright spot* ~0.15 mm

in temperature over periods of seconds which have been shown to persist for hours. The fact that they occur at a wide range of different temperatures appears incompatible with the supposition that they are caused by a liquid–liquid phase transition. Another salient point is that this behaviour is seen at a number of different compositions, not just AY20.

From the recorded video during this pyrometry (Fig. 5.10) it is apparent that the behaviour occurs due to a small significantly brighter (and so, assuming constant emissivity, hotter) point on the sample that rotates around the sample in a repeated pattern. As the polyamorphic rotor operates on the basis of the whole upper third of the sample being at a higher temperature until it experiences a density change, this observation is also inconsistent with the polyamorphic rotor. It is also apparent that the motion of the spot is not a toppling motion, as described by Greaves et al. [5]. The magnitude of the temperature difference in the pyrometry will depend upon the relative proximity of the hot point at its closest approach to the point at which the pyrometer is focussed upon the sample. The period of the oscillation is sensitive to the flow rate, so the exact behaviour recorded by the pyrometer is highly dependent on the experimental conditions. It is seems possible that the hot point is a bubble of trapped gas, although all that can be reasonably concluded is that it is unlikely that pyrometry of this type of behaviour can be attributed to a 'polyamorphic rotor'.

5.5 Conclusions

The controversial issue of the liquid–liquid phase transition reported [5] in AY20 has been studied with small angle neutron scattering. At no temperature or composition was a rise in the SANS intensity consistent with the nucleation of a second liquid phase with a 4 % density difference observed. Systematic errors present in both the SANS measurements presented here and the SAXS measurements of Greaves et al. [5] were considered in order to rectify the observational inconsistency. Systematic pyrometric studies observed very similar behaviour to that of a 'polyamorphic rotor' and, through video recording, associated this with the movements of a small (d ~ 0.15 mm) point on the sample surface. Observations of the same behaviour at a range of temperatures and different sample compositions are inconsistent with that of a 'polyamorphic rotor'.

References

1. Aasland S, McMillan PF (1994) Density-driven liquid–liquid phase separation in the system Al_2O_3–Y_2O_3. Nature 369:633–636
2. Arai Y, Paradis P-F, Aoyama T, Ishikawa T (2003) An aerodynamic levitation system for drop tube and quenching experiments. Rev Sci Instrum 74(2):1057
3. Barnes AC, Skinner LB, Salmon PS, Bytchkov A, Pozdnyakova I, Farmer TO, Fischer HE (2009) Liquid–liquid phase transition in supercooled yttria-alumina. Phys Rev Lett 103(22):225702
4. Dewhurst CD (2003) ILL, Grenoble, France, Internal report, No. ILL03DE01T
5. Greaves GN, Wilding MC, Fearn S, Langstaff D, Kargl F, Cox S, Van QVu, Majérus O, Benmore CJ, Weber R, Martin CM, Hennet L (2008) Detection of first-order liquid/liquid phase transitions in yttrium oxide-aluminum oxide melts. Science 322:566–570
6. Guinier A (1956) X-ray diffraction in crystals, imperfect crystals, and amorphous bodies. Lorrain P, Lorrain DS-M (1994) (trans: French). Dover Publications, Inc., New York
7. Lindner P, Zemb Th (eds) (2002) Neutrons, X-rays and light: scattering methods applied to soft condensed matter. North-Holland, Amsterdam
8. Nagashio K, Kuribayashi K (2002) Spherical yttrium aluminum garnet embedded in a glass matrix. J Am Ceram Soc 85(9):2353–2358
9. Paradis P-F, Yu J, Ishikawa T, Aoyama T, Yoda S, Weber JKR (2003) Contactless density measurement of superheated and undercooled liquid $Y_3Al_5O_{12}$. J Cryst Growth 249(3–4):523–530
10. Skinner LB (2008) Structure and phase behaviour of Aluminate glasses, produced using levitation and laser heating. Ph. D., University of Bristol
11. Skinner LB, Barnes AC, Salmon PS, Crichton WA (2008) Phase separation, crystallization and polyamorphism in the Y_2O_3–Al_2O_3 system. J Phys: Condens Matter 20:205103

Chapter 6
Liquid Invar

6.1 Introduction

The Invar effect, which was discovered by Guillaume [14] at the end of the 19th century, refers to the anomalously low coefficient of thermal expansion of certain alloys [23, 36], $Fe_{65}Ni_{35}$ being the archetype. Different compositions of these alloys also show related effects, such as a constant modulus of elasticity with temperature [12].

Although it has long been realised [34] that these effects were of magnetic origin, the actual mechanism through which this occurs has been highly debated [22, 26, 30]. While a full description of this complex problem is beyond the scope of this work, the essential competing theories are discussed below.

The two main types of model [26] for explaining the Invar effect are those [34] based on a high local magnetic moment-low local magnetic moment transition (HM/LM), and those involving a frustrated local magnetic order [23, 27].

The underlying principle of frustrated local magnetic order models is that it depends on the proposed competition between the positive exchange interactions of Ni–Ni and Ni–Fe atoms, and the negative exchange interaction [15] of Fe–Fe atoms. Exchange interactions, $J[AA]$, which are an essential part of ferromagnetic theories such as the Heisenberg model [19], arise naturally from consideration of the Pauli exclusion principle applied to the potential and kinetic terms of the Hamiltonian. In the case of electrons with a positive exchange interaction the potential term is more significant, and like-like spins for nearest neighbours are favoured (ferromagnetism).

In the case of Invar it has been suggested [27] that the different competing exchange interactions lead to unsatisfied Fe–Fe magnetic bonds, as they are forced by the neighbouring arrangement to adopt like spins. Assuming a simple Ising model, a Leonard-Jones potential for the chemical interactions, and the assumption of equal atomic separations regardless of the pair, Rancourt and Dang [27] determined a relationship for the atomic separations at $T = 0$ K:

$$r \propto r_0^2 J_0'[FeFe]\left[2f_{uFe} - 1\right] + r_0 \tag{6.1}$$

T. Farmer, *Structural Studies of Liquids and Glasses Using Aerodynamic Levitation*,
Springer Theses, DOI: 10.1007/978-3-319-06575-5_6,
© Springer International Publishing Switzerland 2015

where r_0 is the equilibrium separation due purely to chemical interactions, f_{uFe} is the fraction of unsatisfied Fe–Fe magnetic bonds, and:

$$J_0'[FeFe] = \frac{\partial J[FeFe](r_0)}{\partial r} \tag{6.2}$$

where the dependence of the partial derivative on r_0 is explicitly stated. It has been suggested [28] that $J_0'[FeFe]$ is large and positive, and so if $f_{uFe} > 0.5$, as proposed by Rancourt and Dang [27], then the magnetovolume effects cause $r > r_0$. This results in the atomic separation not experiencing the normal thermal expansion.

Although the model of Rancourt and Dang [27] does not require noncollinearity, it should be noted that several authors [22, 30] consider this an important result of the frustrated local magnetic order.

By contrast the transition models [3, 8, 34], which originally stem back to the 2-γ model of Weiss [34], are based upon iron atoms which can have two possible electronic structures. Weiss [34] suggested that for Invar, unlike in pure iron, Fe atoms are in the γ_2 ferromagnetic, HM, high volume state when $T = 0$ K. As the temperature increases, the γ_1 antiferromagnetic, LM, low volume state can be thermally excited. This results in a decrease in atomic separation that balances out the normal thermal increase. Weiss [34] also highlights that the concept of two different magnetic moment states explains several other anomalies that occur in iron and iron alloy.

Crystalline Invar is face centred cubic (FCC) with a lattice constant of 3.59 Å [18]. This corresponds to a nearest neighbour distance of ~2.54 Å. For comparison γ-iron, the FCC allotrope of iron, which exists between 1,667 K and 1,185 K at atmospheric pressure, has a lattice constant [7] of 3.58 Å.

Although there have been a number of different attempts at experimentally determining the mechanism behind the Invar effect, the problem is yet to be conclusively resolved [21, 22]. A consideration of how the models extend into the liquid state, particularly with respect to magnetic correlations, may allow for certain possibilities to be excluded.

6.2 Methods

Samples of $^{Nat}Fe_{65}{}^{Nat}Ni_{35}$, $^{Nat}Fe_{65}{}^{58}Ni_{35}$, and $^{Nat}Fe_{65}{}^{60}Ni_{35}$ were produced from ^{Nat}Fe (99.98 % pure), ^{Nat}Ni (99.8 % pure), ^{58}Ni (99 % pure) and ^{60}Ni (>95 % pure) powders using the two ball method described in Sect. 3.1.2. A pure ^{Nat}Fe sample was also fused for comparison of both the short range atomic ordering and the small angle magnetic scattering. For clarity the compositional denotation will be excluded hereafter (e.g. $^{Nat}Fe^{Nat}Ni$). The scattering lengths of the elements are shown in Table 6.1, and the effect this had on the relative weightings in each sample is shown in Fig. 6.1. The high melting points (~1,700 K) and large reflectivity ($\varepsilon \sim 0.35$ at ~1 μm) [20] of the samples limited their radii to ~1.2 mm.

Table 6.1 The scattering lengths of the natural iron, natural nickel and the two nickel isotopes [6]

Element	Scattering length (fm)
^{Nat}Fe	9.45 (2)
^{Nat}Ni	10.3 (1)
^{58}Ni	14.4 (1)
^{60}Ni	2.8 (1)

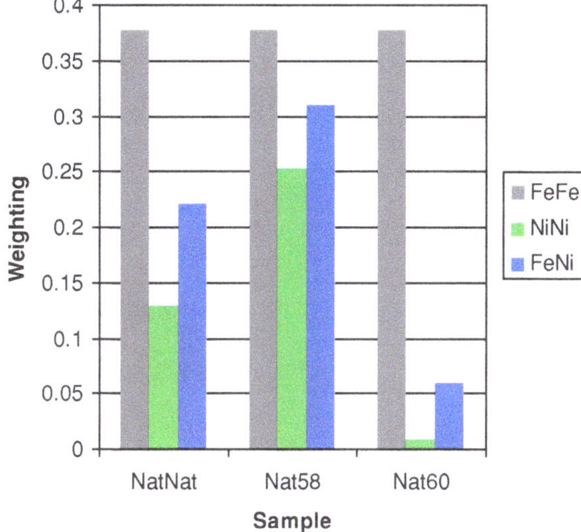

Fig. 6.1 The relative weighting factors of the Fe–Fe, Ni–Ni and Fe–Ni partial pairs based on each sample having 100 % pure components

Table 6.2 Relevant parameters for the neutron diffraction data

Wavelength	0.6961 Å			
Beam height	4 mm			
Beam width	35 mm			
Sample	$^{Nat}Fe_{65}^{Nat}Ni_{35}$	$^{Nat}Fe_{65}^{58}Ni_{35}$	$^{Nat}Fe_{65}^{60}Ni_{35}$	^{Nat}Fe
Initial mass (mg)	49.6	52.6	43.7	44.4
Temperature (K)	1,750	1,820	1,740	1,790
Approximate radius (mm)	1.15	1.20	1.10	1.14
Number density (atom/Å^3)	0.0787			0.074

As with all isotopic substitution experiments the sample number density is assumed to be identical for all samples. The temperatures shown have been corrected for emissivity based upon temperature at which melting occurred

In order to reduce the possibility of sample contamination by oxygen, all laser heating was carried out under an ultrapure (99.9995 %) argon atmosphere. This was achieved by pumping the chamber down to ~100 mTorr, and flushing with argon for 10 min before filling with Ar up to just below atmospheric pressure.

Neutron diffraction was carried out on the samples in the liquid state using in situ aerodynamic levitation on the D4c diffractometer at the ILL [9]. The essential diffractometer setup is described in Table 6.2. The density was estimated by extrapolating between the liquid iron [35] and liquid nickel [2] densities of 7.1 and

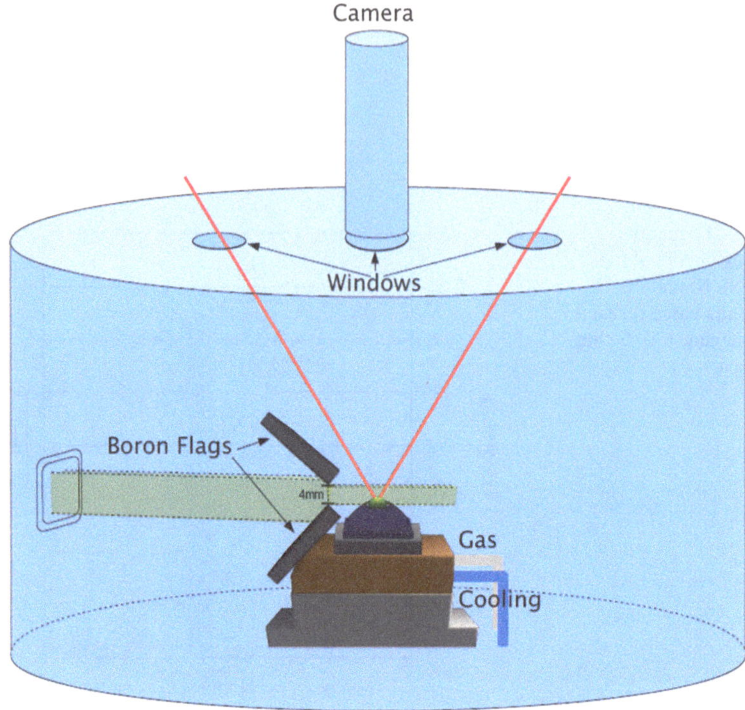

Fig. 6.2 The bell jar setup used during aerodynamic levitation on D4c. The beam profile (shown in *yellow*) has been simplified so that the umbra and penumbra are not shown

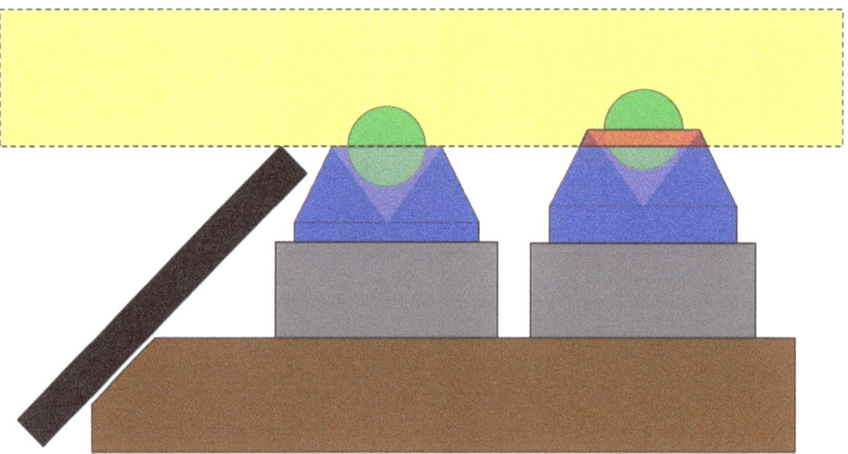

Fig. 6.3 A schematic of the initial vanadium nozzle position, and the position after the vanadium nozzle had risen into the beam

8.05 g cm^{-3}. It was difficult to achieve stable levitation for an extended period due to a small oxide layer that partially covered the surface of the samples. As the partial

Table 6.3 Comparing the theoretical and experimental low *r* limits

Sample	Theoretical low r limit (barns)	Experimental low r limit (barns)
$^{Nat}Fe^{Nat}Ni$	0.95	0.90
$^{Nat}Fe^{58}Ni$	1.25	0.95
$^{Nat}Fe^{60}Ni$	0.50	0.41

oxide layer had a significantly higher emissivity than the metal, this caused a significant increase in temperature if the partial oxide layer was exposed to the laser, which led to sample instability. This oxide layer was sufficiently minimal that there was minimal apparent contribution to the measured diffraction patterns (Fig. 6.6).

The diffractometer/levitator setup is shown in Fig. 6.2. The boron flags were set up in order to prevent the top of the vanadium nozzle from being exposed to the full profile of the beam. A very small amount of the beam's penumbra would have overlapped the top of the vanadium nozzle, contributing a small amount of incoherent scattering to the background.

After the first sample run it was determined that the vanadium nozzle had risen up above the bottom flag during the sample run which had caused an increase in the background scattering (Fig. 6.3). In order to account for this, background scans were taken both before and after every sample scan. Background scans consisted of either a vacuum in the empty belljar, or with a flow of argon typical of that required for levitation. Along with this a nickel scan was taken for calibration and a vanadium sample scan for normalisation.

6.3 Results

Initially the raw sample data had the background subtracted and were corrected for multiple scattering, attenuation and inelastic scattering. A small correction was made for the time evolution of the argon background. Despite taking background scans after the sample scans it was still expected that these would not necessarily be identical to the background at the end of the sample run. This is due to the high probability that in removing the sample the nozzle height was reduced slightly, therefore reducing the amount of incoherent scattering in the background relative to the contribution to the sample scan. Normally this fact could be corrected for, as after all the relevant corrections and normalisation to vanadium the data should match the theoretical limit at high q in the differential scattering cross-section (DSC). However in a levitation experiment the vanadium can only be used as a guide for calibration. This is due to the large uncertainty in how much of both the vanadium and each sample is in the beam. Therefore in this case the data was auto-normalised by fitting to the theoretical high q limit. The validity of this was then checked by comparing the low r value in the $G(r)$ with the theoretical limit calculated from the partial prefactors. Initially there was some disparity between these values in both the $^{Nat}Fe^{58}Ni$ and $^{Nat}Fe^{60}Ni$, as shown in Table 6.3.

Table 6.4 Comparing the normalisation factor determined with a vanadium sample from that required to meet the high q limit

Sample	Factor required to meet high q limit	Vanadium normalisation factor
$^{Nat}Fe^{Nat}Ni$	5.217	5.19
$^{Nat}Fe^{58}Ni$	2.52	4.57
$^{Nat}Fe^{60}Ni$	4.973	5.93
^{Nat}Fe	5.25	5.60

These differences can arise because of one of four things:

- **The number density is too high, reducing the magnitude of the $G(r)$.**
  - As the number density is only estimated (as discussed above) it is probable that it is not completely accurate. However in order to account for the differences the number density would have to be reduced by ~25 %, which is very improbable. Combined with this is the fact that this would in turn cause the $^{Nat}Fe^{Nat}Ni$ low r limit to be significantly different from the theoretical value.
- **One or all of the multiple scattering, Placzek, and attenuation corrections is erroneous.**
  - Considering the nature of the samples none of these effects should have a significant impact on the data, and so it is unlikely that a miscalculated correction is responsible for the discrepancies.
- **The two nickel isotopes are not completely pure.**
  - It is unlikely that the nickel isotopes are completely pure, and so this will affect the theoretical low r limit. However it will also affect the high q limit to which the DSC is being fitted, and the two effects act to partially cancel. Furthermore as the samples are only 35 % nickel any impurity in the nickel isotopes has a reduced impact upon each Invar sample.
- **The amount of incoherent background removed from the sample is too small.**
  - Despite the fact that background scans were taken after each sample run was completed, the possibility that the incoherent background was actually larger during the sample scan is the most probable explanation for the inconsistency with the theoretical low r limit.

Therefore each data set had an amount of incoherent background removed from it which caused the experimental low r limit to coincide with the theoretical limit, having fixed the high q limit to the theoretical value. For the $^{Nat}Fe^{58}Ni$ and $^{Nat}Fe^{60}Ni$ this corresponded to the vanadium nozzle being approximately 1 and 0.3 mm higher in the sample scan than the following background respectively. This was consistent with measurements of nozzle rises taken after scans.

In order to check if the factor required to fit the DSC to the high q limit was of the correct order of magnitude, the vanadium normalisation was calculated. This was based upon exactly half of both the vanadium sample and each Invar sample being in the beam. The $^{Nat}Fe^{Nat}Ni$ and $^{Nat}Fe^{60}Ni$ factors required to meet the high q limit are relatively similar to the calculated vanadium normalisation factors (shown in Table 6.4).

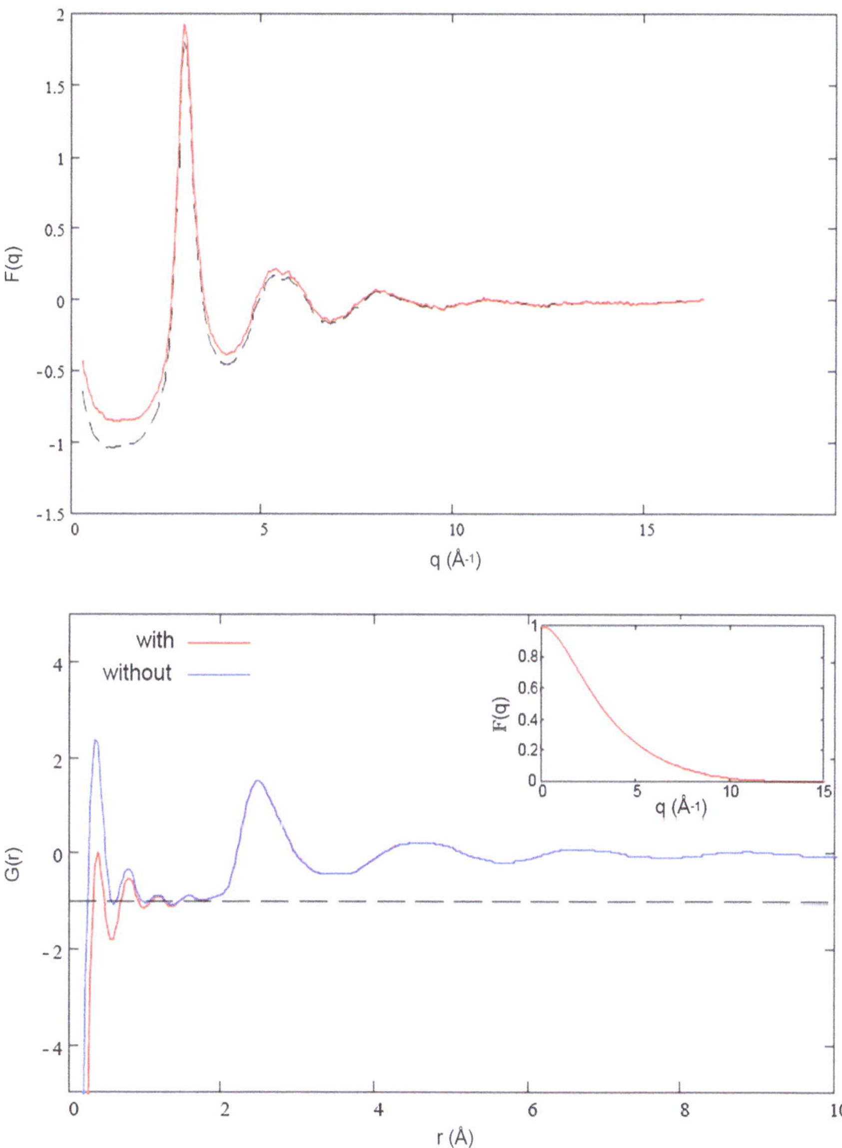

Fig. 6.4 *Top* the $^{Nat}Fe^{Nat}Ni$ total structure factor with (*red solid line*) and without (*broken black line*) a paramagnetic correction. *Bottom* the prefactor normalised total pair correlation function for with and without a paramagnetic correction. The *inset* shows the magnetic form factor for iron

However due to the spherical sample geometry, if the beam instead passed through a point 0.1 mm above the centre of the sample the volume of illuminated sample reduces by 12.5 % for the largest sample. Despite this consideration the $^{Nat}Fe^{58}Ni$ factor is still anomalously low, unless the effects of the nozzle rising are taken into consideration (Fig. 6.3). As the $^{Nat}Fe^{58}Ni$ had the greatest rise in the

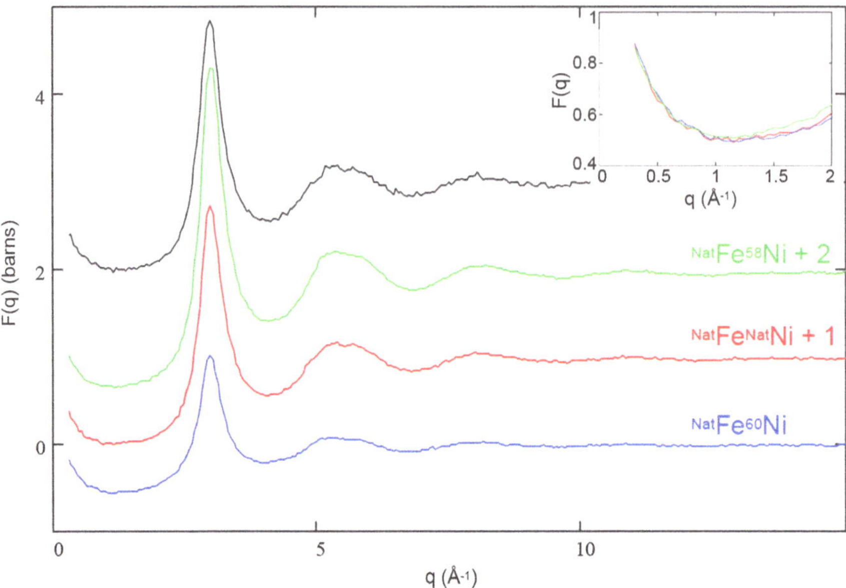

Fig. 6.5 The $F(q)$ data for $^{Nat}Fe^{Nat}Ni$ (*red*), $^{Nat}Fe^{58}Ni$ (*green*), and $^{Nat}Fe^{60}Ni$ (*blue*), and the $S(q)$ for ^{Nat}Fe (*black*). The $^{Nat}Fe^{58}Ni$, $^{Nat}Fe^{60}Ni$, and ^{Nat}Fe data have been shifted up for clarity. The ^{Nat}Fe data also extends to 16.55 $Å^{-1}$ but is covered by the Inset. The three Invar $F(q)$ and the Fe $S(q)$ show little apparent difference. *Inset* the small angle rise due to magnetic scattering is compared for the three Invar samples and found to be in good agreement within experimental errors

nozzle height it is likely that a large extra volume was illuminated by the beam. This would explain why the vanadium normalisation factor, which was determined on the assumption of a hemisphere in the beam, is significantly different from the required factor.

Another consideration was the effect of paramagnetic scattering. Based on the neutron diffraction analyses of Ni [32] and Fe [31] it was estimated that only the paramagnetic scattering from Fe needed to be accounted for. Following the method of Waseda and Suzuki [31], the free atom paramagnetic scattering for Fe was calculated in a proportion sufficient to minimise the low r oscillations in the $G(r)$, as shown in Fig. 6.4. The free paramagnetic scattering factors discussed in Sect. 3.2.6 were calculated from values determined by Freeman and Watson [11].

It should also be mentioned that not only were the three sample temperatures different by up to 80 K (Table 6.2), the temperature of each run fluctuated by as much as ± 50 K. Although it is possible that both of these factors might lead to significant variations in the structure which would invalidate the isotopic substitution, the diffraction measurements on iron of Holland-Moritz et al. [17] suggest that the structure will have remained constant within experimental precision.

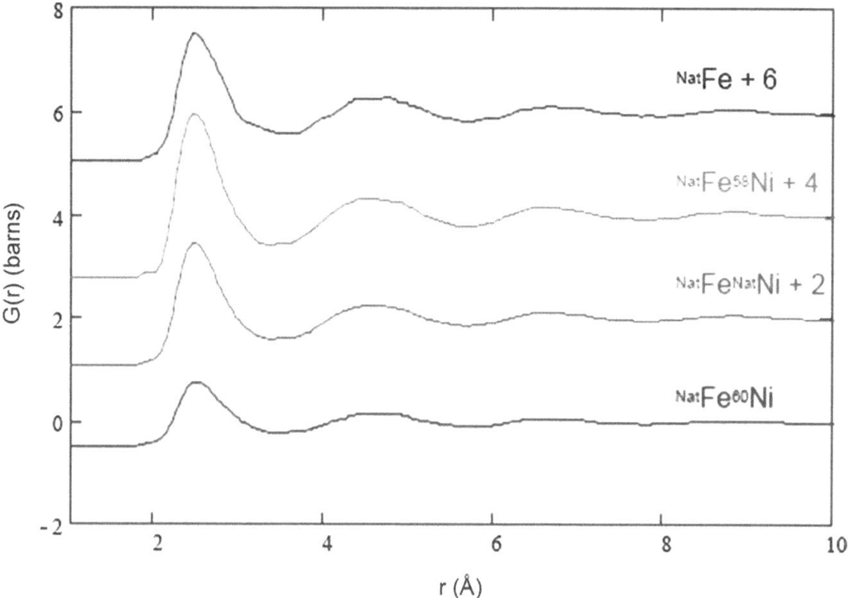

Fig. 6.6 The *G(r)*'s for $^{Nat}Fe^{Nat}Ni$ (*red*), $^{Nat}Fe^{58}Ni$ (*green*), and $^{Nat}Fe^{60}Ni$ (*blue*), and the *g*(*r*) for ^{Nat}Fe (*black*). The $^{Nat}Fe^{58}Ni$, $^{Nat}Fe^{60}Ni$, and ^{Nat}Fe data have been shifted up for clarity. The low *r* oscillations have been cutoff below a distance of 1.8 Å. Again, the Invar and iron data shows little difference. The feature that is visible in the $^{Nat}Fe^{58}Ni$ data at ~1.9 Å suggests a minute amount of oxygen contamination is present, which is not uncommon in the literature [5]. This is reflected in the difference in the back Fourier transforms (Fig 6.7)

The *F*(*q*) for the three Invar samples is shown in Fig. 6.5. The first peak position in all three *F*(*q*)'s is at 3 $Å^{-1}$. The small angle scattering, which is purely from magnetic contributions, is in good agreement in all three samples. This is a good indication that the data has been correctly normalised.

The matrix of the relative weighting factors shown in Fig. 6.1 was inverted (as described in Sect. 3.2.3) in order to determine the Faber-Ziman (FZ) partial structure factors:

$$\begin{bmatrix} S_{NiNi}(q) \\ S_{FeFe}(q) \\ S_{NiFe}(q) \end{bmatrix} = \begin{bmatrix} -26.547 & 9.383 & 17.164 \\ -3.475 & 4.518 & 1.607 \\ 13.009 & -6.603 & -6.406 \end{bmatrix} \begin{bmatrix} F_{NiNat}(q) \\ F_{Ni58}(q) \\ F_{Ni60}(q) \end{bmatrix}$$

The conditioning of the matrix, as discussed in Sect. 3.2.3, is 0.039. Along with the determination of the FZ partial structure factors, the Bhatia Thornton partial structure factors were also computed, resulting in a matrix with a conditioning of 0.314. The FZ partial structure factors and partial pair correlation functions are shown in Fig. 6.8, while the BT partial pairs are shown in Fig. 6.9. The FZ partial structure factors were found to obey the consistency checks discussed by Fischer et al. [10].

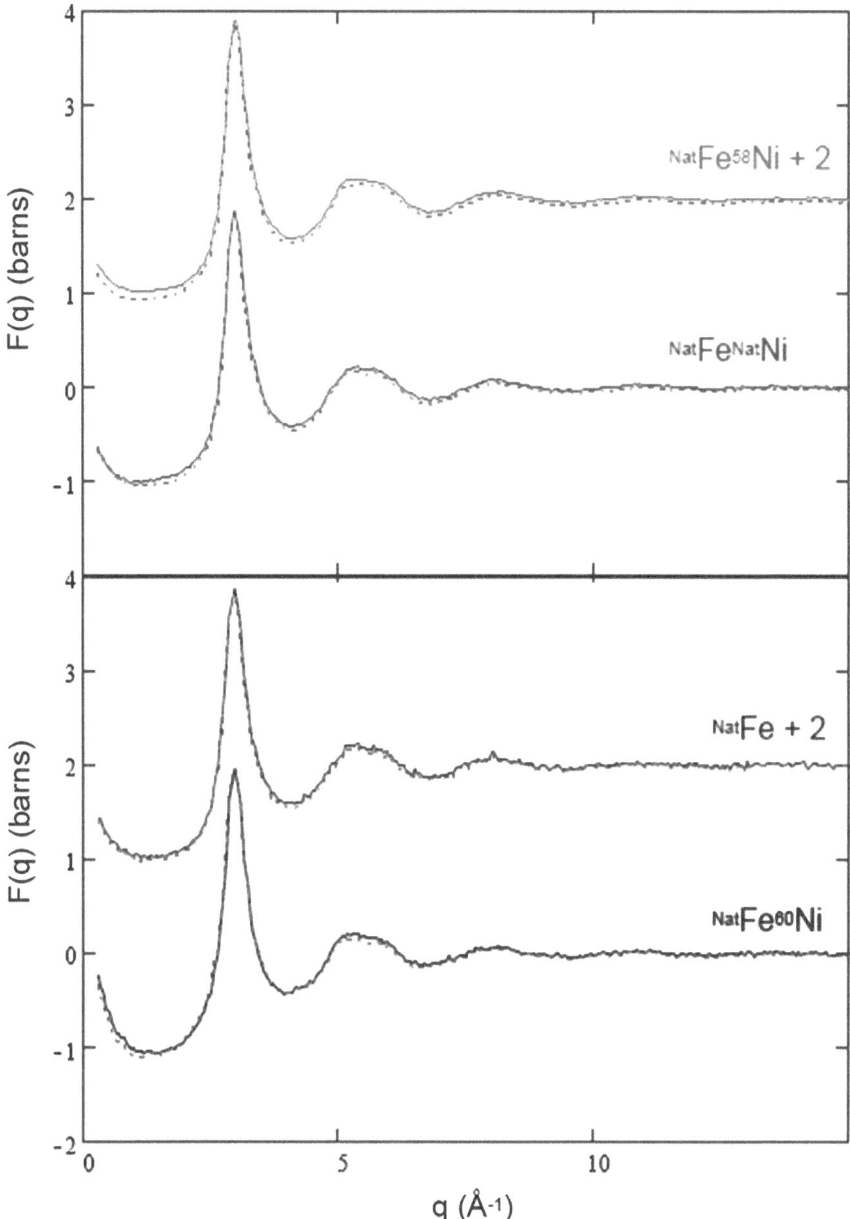

Fig. 6.7 The back Fourier transforms (*solid lines*) of the cutoff $G(r)$ (and $g(r)$ for iron) shown in Fig. 6.6 compared to the original data (*dotted lines*)

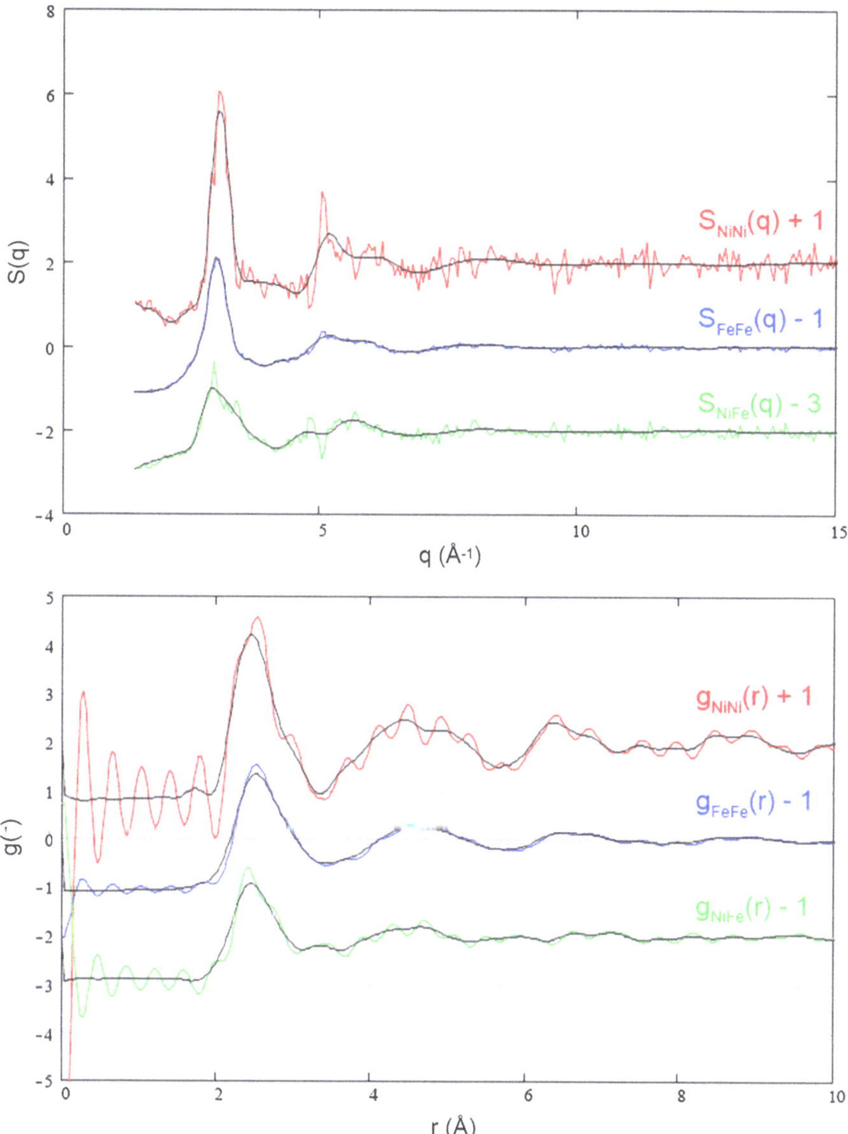

Fig. 6.8 The Faber-Ziman partial structure factors and partial pair correlation functions for $Fe_{65}Ni_{35}$. The *black lines* in the partial structure factors are spline fits to the data. The *black lines* in the partial pair correlation functions are Lorch modified Fourier transforms of the same data. The poorer conditioning of the FZ matrix results in noisier partial structure factors than for the BT formalism. Despite this there appears to be relatively little difference between the Ni–Ni, Fe–Fe, and Ni–Fe partial structure factors, suggesting that the mixing is close to that of an ideal liquid. The very sharp peak observed in $S_{NiNi}(q)$ [with corresponding dip in $S_{NiFe}(q)$] appears to be anomalous, although the origin of this has not been determined

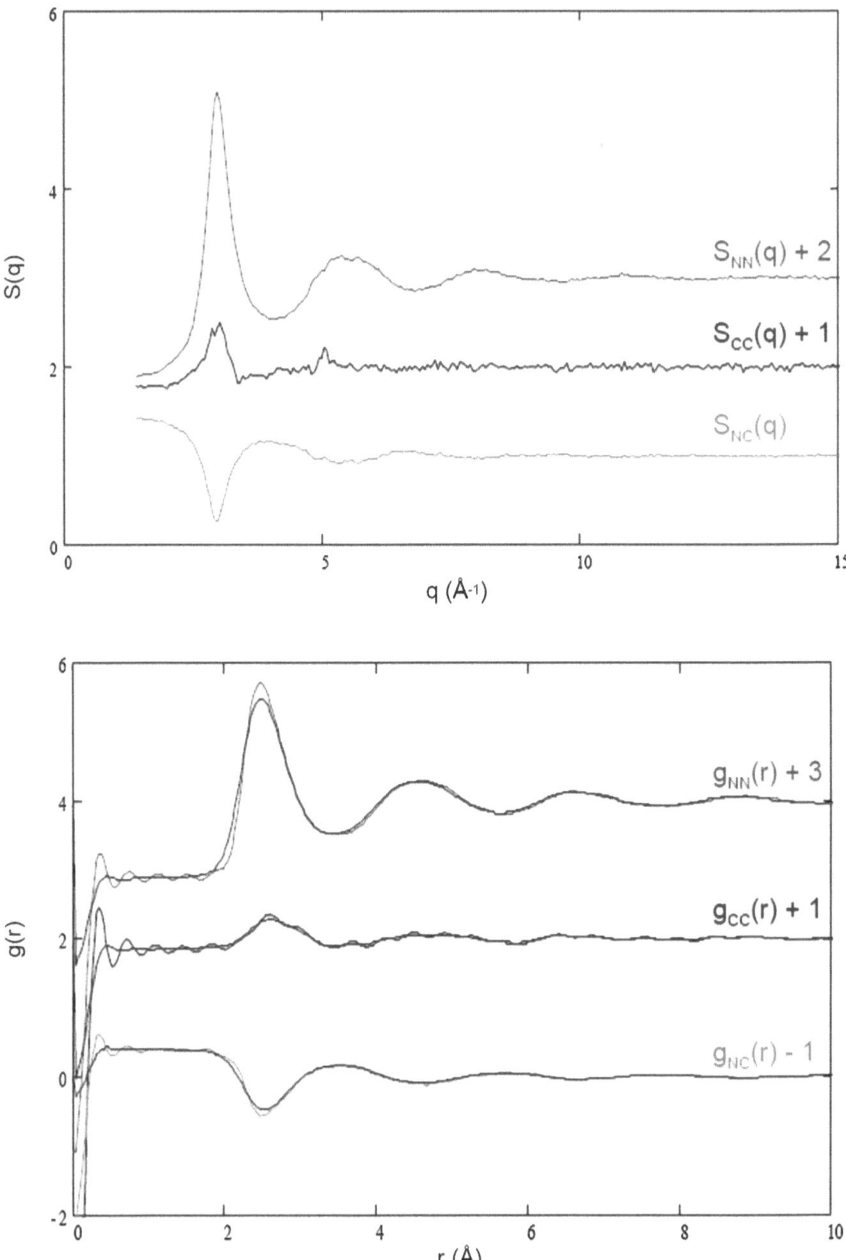

Fig. 6.9 The Bhatia–Thornton partial structure factors and partial pair correlation functions for $Fe_{65}Ni_{35}$. The better conditioned BT partials, specifically $g_{cc}(r)$, suggest that while the distributions of Fe and Ni atoms in $Fe_{65}Ni_{35}$ is broadly similar, there is degree of ordering present

Table 6.5 The first peak positions and coordination numbers for Fe determined in this work compared with two previous neutron diffraction studies

	First peak position (Å)	Coordination number
Waseda and Suzuki	2.58 [RDF]	9–10.3
Holland-Moritz et al.	2.55 [$g(r)$]	12.3
This work	2.58 ± 0.02 [RDF] 2.50 ± 0.02 [$g(r)$]	12.8 ± 0.6

In the case of Waseda and Suzuki [31] the first peak position is only given for the radial distribution function, as indicated by the brackets. The temperatures used in each study are as follows: 1,890 K for Waseda and Suzuki [31]; 1,870 K for Holland-Moritz et al. [17]; 1,790 K for this work

6.4 Discussion

Neutron and X-ray diffraction measurements have previously been made on pure iron by Waseda and Suzuki [31] using alumina containers, and by Holland-Moritz et al. [17] using electromagnetic levitation. Although there is broad agreement between the structure factors of each group, there are significant differences in the pair distribution functions, which are highlighted in Table 6.5. The most important difference is in the coordination number. Waseda and Suzuki [31] suggested that there was not a significant change in the coordination number during melting from the BCC δ-Fe. Holland-Moritz et al. [17], by contrast, suggested that the coordination number is closer to that of the lower temperature stable allotrope, γ-Fe (FCC). Along with this there is a slight disagreement over the first peak position.

The structure factor and pair distribution function determined for Fe are shown in Figs. 6.5 and 6.6. The coordination number agrees with that of Holland-Moritz et al. [17], suggesting that there is a large change in the coordination upon melting. The first peak position however is closer to that of Waseda and Suzuki [31], although it is important to note that they performed measurements at a temperature ~100 K above that used in this work. For comparison, in γ-Fe (1,183–1,673 K) and δ-Fe (1,673–1,808 K) [25] the nearest neighbour distances are 2.53 and 2.54 Å respectively (Fig. 6.7).

In considering the structure of Invar, it is illustrative to consider the extent to which the introduction of nickel affects the pure iron structure. Figure 6.5 compares the pure Fe and the Invar data, and shows that the structures are very similar. The peak positions also compare well with the peak position of 2.50 Å of $Fe_{85}Ni_{15}$ reported by Cuello et al. [5]. This indicates that the Ni effectively replaces some of the Fe atoms. This possibility, which is also suggested by the coordination numbers in Tables 6.6 and 6.7, might be expected based on their similar chemistries.

The implication that the preferential atomic ordering is the same for the Ni and Fe atoms is furthered by the total structure factors, shown in Fig. 6.10. Here the total structure factors from the three different samples are normalised by relevant scattering lengths, resulting in them being almost identical. Further confirmation of this can be determined from a comparison of the partial structure factors and

Table 6.6 The first peak position and total coordination number of the three Invar samples compared with iron

	First peak position (Å)	Coordination number
$^{Nat}Fe^{Nat}Ni$	2.49	13.3
$^{Nat}Fe^{60}Ni$	2.50	13.7
$^{Nat}Fe^{58}Ni$	2.49	13.0
^{Nat}Fe	2.50	12.8

The similar coordination number and first peak position in the Invar samples suggests that the Ni atoms effectively substitute into the pure Fe structure. The uncertainty on the first peak positions and coordination numbers is 0.02 and 0.6 respectively. The relatively large error in the coordination number occurs in part due to the poorly defined first minimum in each $G(r)$

Table 6.7 Coordination numbers determined from the FZ partial pair correlation functions shown in Fig. 6.8

	Coordination number
Fe around Fe	8.8 ± 0.6
Ni around Ni	4.6 ± 0.8
Fe around Ni	8.3 ± 0.8
Ni around Fe	4.5 ± 0.8

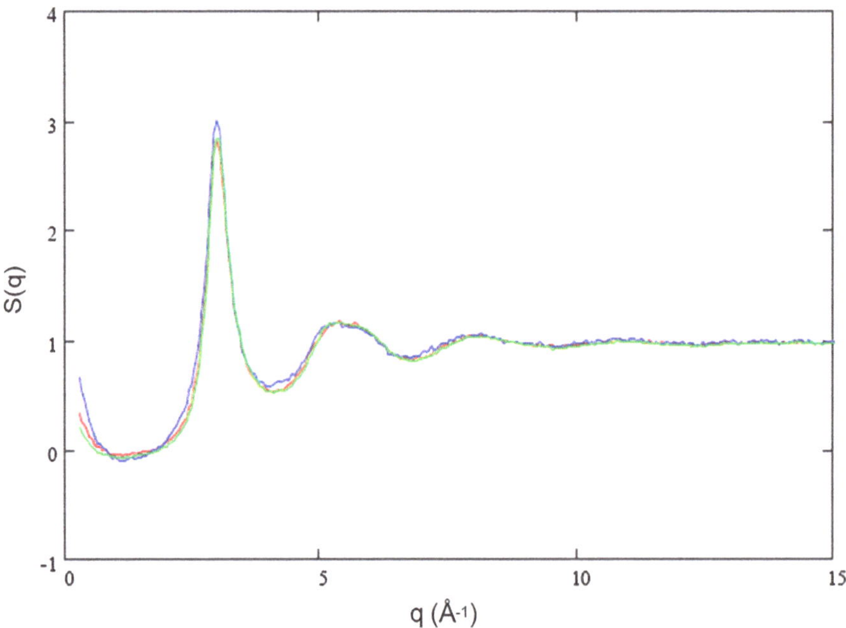

Fig. 6.10 The total structure factors for $^{Nat}Fe^{Nat}Ni$ (*red*), $^{Nat}Fe^{58}Ni$ (*green*), and $^{Nat}Fe^{60}Ni$ (*blue*). The small angle region should be discounted, as the data is being normalised to the nuclear scattering and the small angle region is largely due to magnetic scattering

the partial pair correlation functions (Figs. 6.8 and 6.9). It is appears from the FZ partial pair correlation functions that the ordering in the Ni–Ni, Fe–Fe, and Ni–Fe partials is similar. However the better conditioned BT C–C partial suggests that there is a small degree of preferential ordering (Fig. 6.9).

As well as the atomic ordering, the good low q range accessible on D4c allows some consideration of the small angle scattering. In the case of a neutron scattering arising purely from neutron-nuclei interactions in a monatomic system, the following limit exists:

$$S(0) = \rho_0 \chi_T k_B T \tag{6.3}$$

where χ_T is the isothermal compressibility, k_B is Botlzmann's constant, and T is the absolute temperature [10]. For liquid iron at 1,790 K using compressibility data determined by Hixson et al. [16], this equates to a minimum of 22.3 mbarns. From the rise in the small angle scattering in both the pure Fe and $Fe_{65}Ni_{35}$ data it is obvious that the largest contribution must come from magnetic scattering. Following Van Hove [29, 33] it is apparent that the rise in the small angle scattering originates from spin correlations, where the differential scattering cross section is described by:

$$\left(\frac{d\sigma}{d\Omega}\right) = b^2 I(q) + \frac{2}{3}\left(\frac{e^2\gamma}{mc^2}\right)\mathrm{F}^2(q)S(S+1)\frac{\xi^2}{\eta^2\left(1+\xi^2q^2\right)} \tag{6.4}$$

where $I(q)$ is the atomic structure factor, e is the electron charge, γ is the neutron magnetic moment, m is the electron mass, c is the velocity of light, S is the spin quantum number, ξ is the correlation length, and η describes the strength of the correlation.

Therefore after subtracting the nuclear scattering, it is possible to determine the magnetic correlation length from the inverse of the DSC plotted against q^2, also known as the Ornstein Zernike plot. For iron the Ornstein–Zernike (OZ) plot is shown in Fig. 6.11, which leads to a value of $\xi \approx 1.5$ Å. In Fig. 6.12 this correlation length is compared with the value of Weber, Knoll and Steeb [33], and with a linear extrapolation from measurements [24] taken in the solid but above the Curie temperature, T_C. It is obvious that there is a significant difference from both the previous data and the extrapolated data. However the disparity between the correlation length of Weber et al. [33] and the extrapolated data, which is even more pronounced in their cobalt data, shows the difficulty in accurately determining magnetic correlation lengths in the liquid state.

A similar rise in the small angle scattering is also seen in the three $^{Nat}Fe^{Nat}Ni$ samples which also shows OZ behaviour (Fig. 6.11). The correlation lengths are shown in Table 6.8. Although, based on the above discussion of iron, some caution should be used in quantitative interpretation, it is clear that there are still significant magnetic correlations, possibly even greater in size than in iron. This is somewhat unexpected, as the temperatures at which these samples were measured are very far from the Invar Curie temperature [4] of ~500 K. A comparable situation is seen in Ni where the Curie temperature is 631 K [33]. However

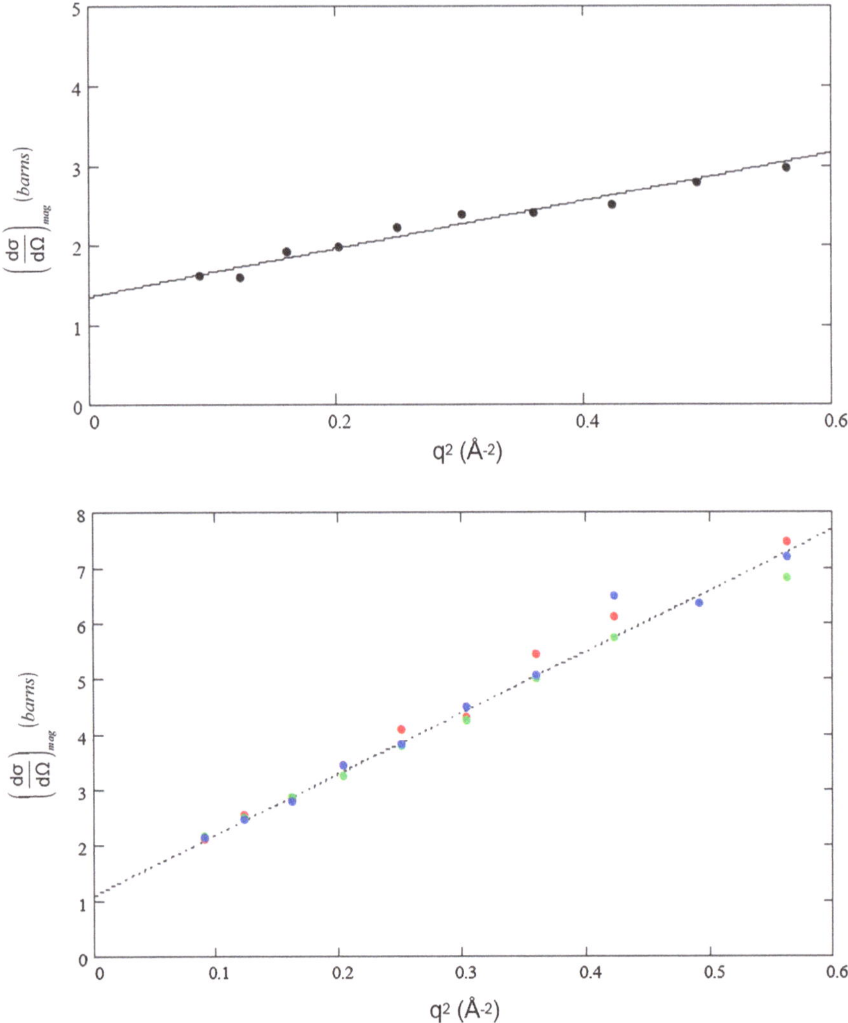

Fig. 6.11 OZ plots for iron (*black circles*), and $^{Nat}Fe^{Nat}Ni$ (*red circles*), $^{Nat}Fe^{58}Ni$ (*green circles*), and $^{Nat}Fe^{60}Ni$ (*blue circles*) . The data obeys a linear relationship up to a minimum of 0.6 $Å^{-2}$, suggesting that OZ behaviour occurs

Weber et al. [33] demonstrated that the magnetic correlations in Ni at 1,773 K have almost disappeared. It would be useful to compare the calculated Invar correlation lengths with Invar correlation lengths close to the Curie temperature; unfortunately there is an apparent lack of experimentally determined magnetic correlation lengths. A comparison can be made with correlation length data for $Fe_{75}Ni_{25}$ from Grigoriev et al. [13]. Extrapolating to a temperature of 1,500 °C yields a correlation length of 2.0 Å, which is similar to those shown in Table 6.8.

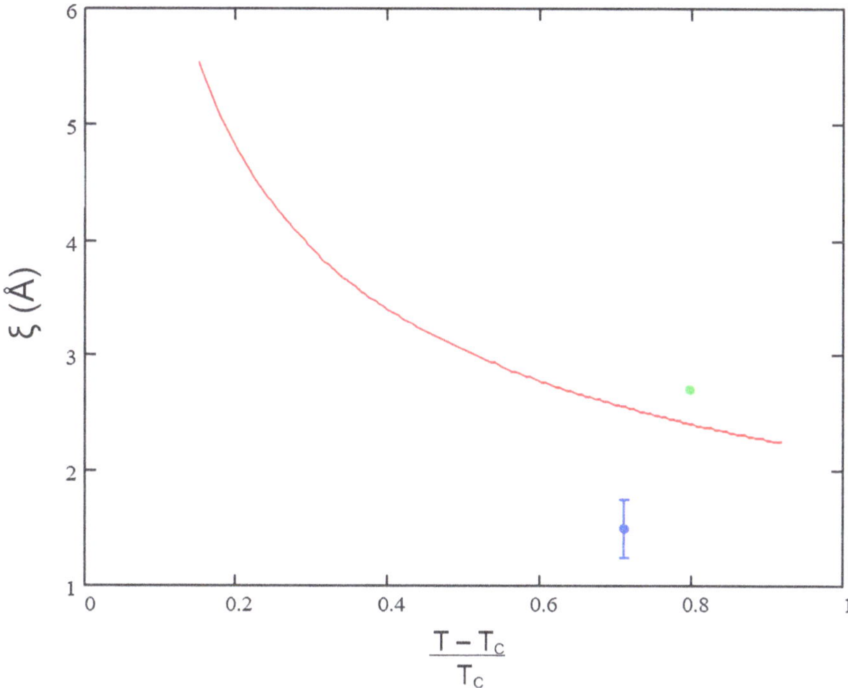

Fig. 6.12 Following Weber et al. [33] the crystalline Fe correlation length data is extrapolated into the liquid region. The *green* cross was the correlation length of 2.7 Å determined by Weber et al. [33], while the *blue dot* is from the current work

Table 6.8 The correlation lengths of $^{Nat}Fe^{Nat}Ni$, $^{Nat}Fe^{60}Ni$, and $^{Nat}Fe^{58}Ni$ determined by fitting the points in the OZ plot (Fig. 6.11)

Sample	Temperature (°C)	Correlation length (Å)
$^{Nat}Fe^{Nat}Ni$	1,480	2.9
$^{Nat}Fe^{60}Ni$	1,470	3.2
$^{Nat}Fe^{58}Ni$	1,550	2.9

The uncertainty in the correlation lengths is 0.3 Å

A certain amount of care should be taken with this comparison, as the $Fe_{75}Ni_{25}$ composition actually has an anomalously large coefficient of thermal expansion [1], also known as the anti-Invar effect. However as these effects almost certainly have a related magnetic cause [1] it is not unreasonable to expect similar magnetic correlation length behaviour. It should again be reiterated that the emphasis should be placed upon the observation of significant magnetic correlations, rather than the quantitative determination of correlations lengths. The existence of significant magnetic correlations in Invar at temperatures far above the Curie temperature might allow crystalline magnetic models to be constrained.

6.5 Conclusions

Both the atomic structure and the magnetic correlations of Invar and Fe were studied using in situ aerodynamic levitation and neutron diffraction with isotopic substitution. Initial observations at the structure factor level suggested that the nickel effectively randomly substituted for the iron atoms with very little change in the structure. However the $S_{CC}(q)$ BT partial structure factor suggests that there is a small degree of ordering present. The magnetic correlations of Invar and Fe were found to be comparable. This result is somewhat unexpected as sample measurements were taken ~1,300 K above the Invar Curie temperature. This suggests that the magnetic correlations in Invar at the Curie temperature are on a significantly greater length scale than the corresponding situation in Ni.

References

1. Acet M, Schneider T, Zähres H, Wassermann EF, Pepperhoff W (1994) Anti-Invar in Fe–Ni. J Appl Phys 75(10):7015–7017
2. Brillo J, Egry I (2003) Density determination of liquid copper, nickel, and their alloys. Int J Thermophys 24(4):1155–1170
3. Chikazumi S (1979) Invar anomalies. J Magn Magn Mater 10(2–3):113–119
4. Crangle J, Hallam GC (1963) The magnetization of face-centred cubic and body-centred cubic iron + nickel alloys. Proc R Soc A: Math, Phys Eng Sci 272(1348):119–132
5. Cuello G, Fernández-Perea R, Bermejo FJ, Román-Ross G, Campo J (2007) Structure of Fe–Ni and Fe–Ni–S molten alloys by neutron diffraction. J Non Cryst Solids 353(32–40):2987–2992
6. Dianoux A-J, Lander G (eds) (2003) Neutron data booklet, 2nd edn. Old City Publishing, Philadelphia
7. Ehrhart P, Schönfeld B, Ettwig HH, Pepperhoff W (1980) The lattice structure of antiferromagnetic γ-iron. J Magn Magn Mater 22(1):79–85
8. Entel P, Hoffmann E, Mohn P, Schwarz K, Moruzzi VL (1993) First-principles calculations of the instability leading to the Invar effect. Phys Rev B 47(14):8706–8720
9. Fischer HE, Cuello GJ, Palleau P, Feltin D, Barnes AC, Badyal YS, Simonson JM (2002) D4c: a very high precision diffractometer for disordered materials. Appl Phys A 74:S160–S162
10. Fischer HE, Barnes AC, Salmon PS (2006) Neutron and X-ray diffraction studies of liquids and glasses. Rep Prog Phys 69(1):233–299
11. Freeman AJ, Watson RE (1961) Hartree-Fock atomic scattering factors for the neutral atom iron transition series. Acta Crystallogr A 14(3):231–234
12. Fukamichi K, Masumoto T, Kikuchi M (1979) Invar and Elinvar amorphous alloys. IEEE Trans Magn 15(6):1404–1409
13. Grigoriev SV, Maleyev SV, Okorokov AI, Runov VV (1998) Observation of two length scales of magnetic correlations in the Invar Fe_{75} Ni_{25} alloy above Tc by means of small-angle neutron scattering and neutron depolarization. Phys Rev B 58(6):3206–3211
14. Guillaume CE (1919–1920) The anomaly of the nickel-steels. Proc Phys Soc Lond 32:374–404
15. Hatherly M, Hirakawa K, Lowde RD, Mallett JF, Stringfellow MW, Torrie BH (1964) Spin wave energies and exchange parameters in iron-nickel alloys. Proc Phys Soc Lond 84(5371):55–62

16. Hixson RS, Winkler MA, Hodgdon ML (1990) Sound speed and thermophysical properties of liquid iron and nickel. Phys Rev B 42(10):6485–6491
17. Holland-Moritz D, Schenk T, Convert P, Hansen T, Herlach DM (2005) Electromagnetic levitation apparatus for diffraction investigations on the short-range order of undercooled metallic melts. Meas Sci Technol 16(2):372–380
18. Kirschbaum U, Zähres H, Wassermann EF (1995) Invar behavior and martensitic transitions in FeNiPt alloys. J Phys IV 5(C2):111–116
19. Kittel C (2005) Introduction to solid state physics, 8th edn. Wiley, Hoboken
20. Kobatake H, Khosroabadi H, Fukuyama H (2012) Normal spectral emissivity measurement of liquid iron and nickel using electromagnetic levitation in direct current magnetic Field. Metall Mater Trans A 43(7):2466–2472
21. Lomova NV, Shabanova IN, Chirkov AG, Ponomaryov AG (2007) On short-range ordering in Fe–Ni alloys. J Electron Spectrosc Relat Phenom 156–158:401–404
22. Matsumoto K, Maruyama H, Ishimatsu N, Kawamura N, Miztumaki M, Irifune T, Sumiya H (2011) Noncollinear spin structure in Fe–Ni Invar alloy probed by magnetic EXAFS at high pressure. J Phys Soc Jpn 80(2):023709
23. Menshikov AZ (1989) On the Invar problem. Physica B 161(1–3):1–8
24. Passell L, Blinowski K, Brun T, Nielsen P (1965) Critical magnetic scattering of neutrons in iron. Phys Rev 139(6A):A1866–A1876
25. Pillai SO (2005) Solid state physics, 6th edn. New Age International, New Delhi
26. Rancourt DG (2002) Invar behavior in Fe–Ni alloys is predominantly a local moment effect arising from the magnetic exchange interactions between high moments. Phase Transitions 75(1–2):201–209
27. Rancourt D, Dang MZ (1996) Relation between anomalous magnetovolume behavior and magnetic frustration in Invar alloys. Phys Rev B: Condens Matter 54(17):12225–12231
28. Sabiryanov RF, Bose SK, Mryasov ON (1995) Effect of topological disorder on the itinerant magnetism of Fe and Co. Phys Rev B, 51(14):8958–8973
29. Van Hove L (1954) Time-dependent correlations between spins and ferromagnetic crystals. Phys Rev 95(6):1374–1384
30. van Schilfgaarde M, Abrikosov IA, Johansson B (1999) Origin of the Invar effect in iron±nickel alloys. Nature 400(6793):46–49
31. Waseda Y, Suzuki K (1970) Atomic distribution and magnetic moment in liquid iron by neutron diffraction. Physica Status Solidi 39(1):p669
32. Waseda Y, Suzuki K, Tamaki S, Taeuchi S (1970) Neutron diffraction study of nickel in liquid state. Physica Status Solidi 39(1):p181
33. Weber M, Knoll W, Steeb S (1978) Magnetic small-angle scattering from molten elements iron, colbalt and nickel. J Appl Crystallogr 11:638–641
34. Weiss RJ (1963) The origin of the 'Invar' effect. Proc Phys Soc Lond 82(526):281–288
35. Wille G, Millot F, Rifflet JC (2002) Thermophysical properties of containerless liquid iron up to 2,500 K. Int J Thermophys 23(5):1197–1206
36. Wohlfarth EP (1979) Invar behaviour in crystalline and amorphous alloys. J Magn Magn Mater 10(2–3):120–125

Chapter 7
Indium Selenide

7.1 Introduction

7.1.1 Liquid Semiconductors

Semiconducting behaviour occurs in crystalline structures due to the formation of a bandgap separating electron states capable of free diffusion (conduction band) from the highest energy occupied state at $T = 0$ K (Fermi energy) [11]. However in a liquid, semiconducting behaviour can exist even if there is no zero in the density of states, $n(E)$ (i.e. no bandgap). This is due to formation of localized states at E_F, which occur due to the disordered spatial distribution of the potentials in an amorphous structure, as initially explained by Anderson [1]. Therefore, there are three broad types of conduction behaviour in liquids: metallic liquids where the atomic scattering of electrons is weak [18, 27], the electron mean free path is long, and $n(E)$ is large at E_F; semi-metallic (or broad definition semiconductor [3] liquids where there is a deep minimum in $n(E)$ (also known as a pseudogap) but the electron wave functions are extended (Fig. 7.1b); and liquid semiconductors (or narrow definition semiconductors) where E_F lies in a 'conductivity gap' (Fig. 7.1c, d). A conductivity gap [3] is an energy region where there is zero conduction at $T = 0$ K, and can occur either because $n(E) = 0$, or if there is a deep minimum in $n(E)$ where the states are localized.

The conductivity, σ, and the thermopower, S, of liquid semiconductors can be understood in terms of the Kubo-Greenwood [7] equations:

$$\sigma = -\int \sigma(E)\frac{df}{dE}dE \tag{7.1}$$

$$S = -\frac{k_B}{e}\int \frac{\sigma(E)}{\sigma}\frac{E - E_F}{k_BT}\frac{df}{dE}dE \tag{7.2}$$

T. Farmer, *Structural Studies of Liquids and Glasses Using Aerodynamic Levitation*, Springer Theses, DOI: 10.1007/978-3-319-06575-5_7,

© Springer International Publishing Switzerland 2015

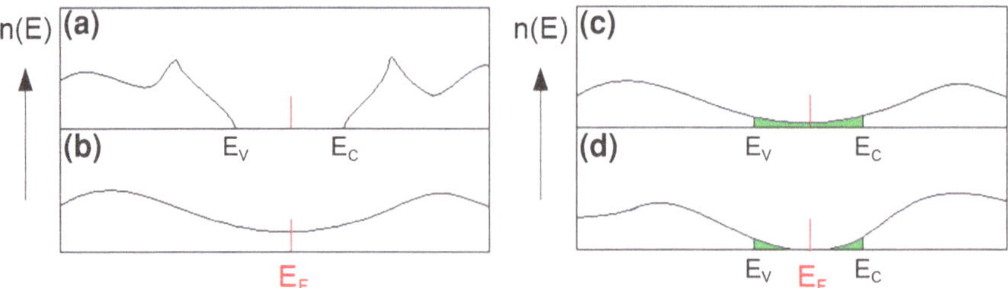

Fig. 7.1 Density of states (*DOS*) for different semiconducting behaviour. **a** Bandgap in a crystalline semiconductor. **b** A semi-metallic liquid where E_F is located in a minimum in the DOS. **c** A narrow definition liquid semiconductor where E_F is located in a minimum consisting of localized states. **d** A narrow definition liquid semiconductor because $n(E) = 0$ where E_F is located

where $\sigma(E)$ is the conductivity at an energy E, f is the Fermi function, E_F is the Fermi energy, T is the absolute temperature, and k_B is Boltzmann's constant. From (7.1) we can attempt to understand the conductivity behaviour with a change in temperature and composition.

An increase in temperature causes the Fermi function to broaden which will lead to an increase in σ. Along with this is the effect of temperature on $\sigma(E)$, which is dependent on $n(E)$. It is likely that the origin of the pseudogap is related to the extent of the heterocoordination in a binary liquid semiconductor [3], and that at stoichiometry E_F is located in this minimum. From this it can be seen that with an increase in temperature the prevalence of heterocoordination will decrease, which will again contribute to an increase in conductivity. Therefore both contributing factors will cause an increase in conductivity as the temperature increases.

From the suggestion that the pseudogap will be deepest at the stoichiometric composition, it is obvious that this should be the composition of minimum conductivity, which is the case for a number of semiconductor liquids [3].

7.1.2 Conductivity Behaviour of Liquid In–Se

In both of the respects discussed above, the liquid In–Se system displays unusual conductivity behaviour, as shown by Okada and Ohno [21] (Fig. 7.2). A possible source of this unusual behaviour is the tendency for In to form compounds of both In^{+} and In^{3+}, as it lies between the principally trivalent Ga [14], and the monovalent Tl [2]. At the stoichiometric composition In_2Se_3 there is a local maximum, which is not consistent with the concept that the heterocoordination should be at a maximum in the conductivity, with bonding occurring between In^{3+} and Se^{2-}. In_2Se_3 also shows only a small increase in conductivity with temperature. The magnitude of this maximum decreases with increasing temperature, and extrapolation suggests [13] that In_2Se_3 might develop a conductivity minimum at ~1,375 K. The only other systems which exhibit similar behaviour [20] are Ag–S and Ag–Se, which both have a local maximum at Ag_2M.

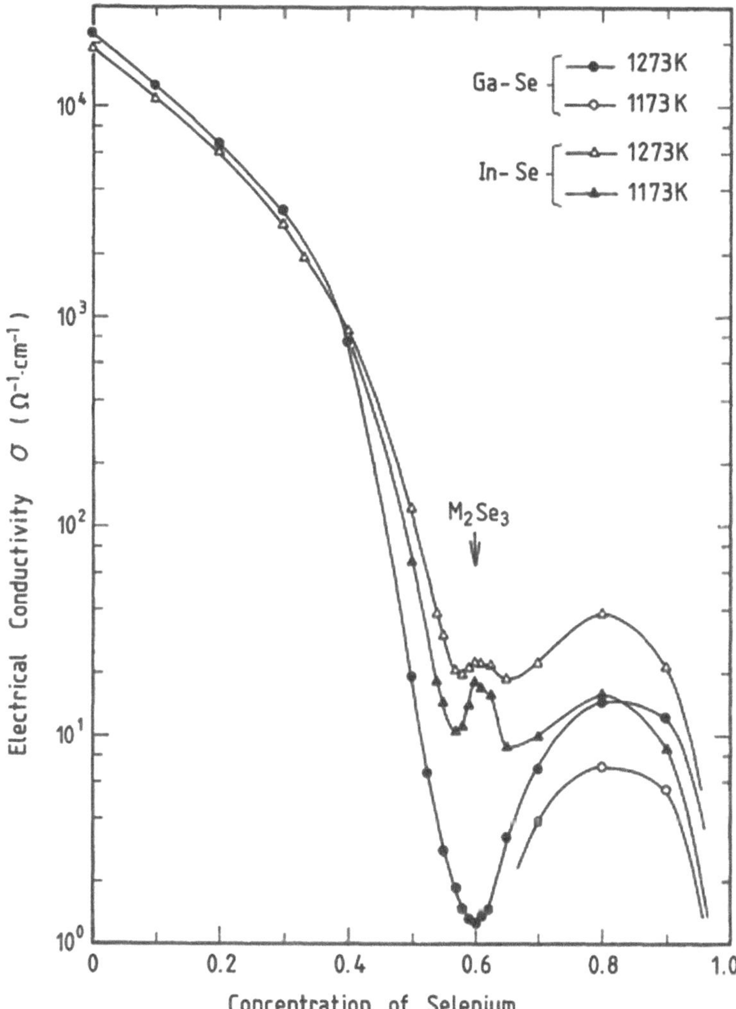

Fig. 7.2 The conductivity behaviour of indium selenide and gallium selenide as determined by Okada and Ohno [22]. The conductivity minimum occurs at a composition between InSe and In_2Se_3, rather than at stoichiometry as expected. The anomalous peak at In_2Se_3 appears to disappear with increasing temperature (Reprinted by permission of Physical Society of Japan. Okada and Ohno [22])

However in these cases the temperature dependence of the conductivity is different ($d\sigma/dT < 0$) so it is not clear the underlying mechanism is the same.

Along with this anomalous maximum there is also the consideration why such a range of compositions display good semiconducting behaviour. On the Se rich side of In_2Se, the conductivity enters a strong scattering regime, as is evidenced by the conductivity dropping below ~3,000 $(\Omega\ \mathrm{cm})^{-1}$ and $d\sigma/dT$ changing from negative to positive (Fig. 7.2). The existence of a relatively high conductivity at

In_2Se suggests that the heterocoordination is not pronounced, which in turn suggests that there is an absence of In^+ ions. Around the InSe composition the conductivity is on the boundary between the broad and narrow definitions of liquids semiconductors, and so E_F is probably located in a very deep pseudogap at the onset of becoming a conductivity gap. It is difficult to explain this behaviour just considering trivalent In and a tendency for heterocoordination.

It is useful to consider the conductivity behaviour and structure of the other Se-Group III semiconducting liquids GaSe [22] (Fig. 7.2) and TlSe [2]. Ga is usually found in a trivalent state, whereas Tl is generally monovalent. This is reflected in the fact that the conductivity minima are found at the stoichiometric compositions Ga_2Se_3 and Tl_2Se, unlike the unusual local maximum at In_2Se_3. However all three exhibit very low conductivity across large compositional ranges, in contrast with what is expected as the composition moves away from stoichiometry. It has been suggested, based on both neutron diffraction [14] and ab initio molecular dynamics [8], that as GaSe moves away from the stoichiometric composition Ga–Ga homopolar bonding occurs. It is possible that this is the same with $In_{1\text{-}x}Se_x$, as posited by Ferlat et al. [4], however the potential for mixed valence states and the occurrence of the anomalous stoichiometric maximum means that it is important the mechanism is established.

Along with the interesting semiconducting behaviour, InSe is also of interest from a technological perspective. Recently there has been growing interest in the possibility of using chalcogenides for phase change memory [10], and InSe has been proposed as a possible candidate [15]. This phase change memory consists of melting transition from a crystalline state to a liquid state, followed by quenching to an amorphous state. Therefore the liquid structure and electrical properties are important not only in establishing the effects of the transition, but also with relation to the resulting amorphous structure.

Further to this is the consideration of the applicability of ANS to aerodynamic levitation mentioned in Chap. 1, which will be considered in Sect. 7.4.

7.1.3 Crystalline Structure

Crystalline InSe has a four layer structure, Se–In–In–Se, with each layer exhibiting a hexagonal symmetry [16]. This is equivalent to tetragonal In atoms covalently bonded to three Se atoms and another In atom. The nearest neighbour interatomic distances are 2.64 Å for In–Se, 2.82 Å for In–In, and 3.86 Å for Se–Se [16]. The band gap at 293 K is 1.19 eV [24].

7.2 Methods

InSe and In_2Se_3 samples were prepared by sealing the correct relative proportions of ^{Nat}In (99.999 % pure) and ^{Nat}Se (99.5 % pure) powders into an evacuated silica tube. The intention was to measure both samples, however due to a heat shielding

Table 7.1 Relevant parameters for the neutron diffraction data. The sample density was taken from Lague [13]

Geometric parameters					
Inner container diameter			0.616 cm		
Outer container diameter			0.820 cm		
Sample height			5 cm		
Beam width			1.4 cm		
Beam height			4.2 cm		
Vanadium diameter			8 mm		
Sample density			5.020 g cm^{-3}		
Wavelength dependent InSe parameters					
λ (Å)	In b_{real} (fm)	In b_{imag} (fm)	InSe σ_{coh} (b)	InSe σ_{inc} (b)	InSe σ_{abs} (b)
0.5	3.49	0.09	4.77	0.625	45.28
0.33	1.89	0.23	4.25	1.28	73.41
0.29	−0.05	0.48	4.07	2.42	134.82

problem the In_2Se_3 sample was not measured and so no further mention of it will be made. The silica tube containing the InSe was heated to 800 °C (~150 °C above the liquidus [5]) in a rocking furnace for 10 h in order to ensure homogenous mixing.

Neutron diffraction was carried out on the liquid samples on the D4c diffractometer [6] at the ILL, using neutron wavelengths of 0.5, 0.33, and 0.29 Å. A Cu(220) monochromator was used for 0.5 Å and a Cu(331) monochromator was used for both the 0.33 and 0.29 Å neutrons. In order to avoid harmonic contamination a rhodium filter was inserted between the monochromator and the sample when measuring at 0.5 Å. Diffraction measurements were taken at the three wavelengths on the empty furnace, the empty container, the liquid sample, nickel powder in a vanadium can for calibration, a vanadium bar for normalisation, and ^{10}B powder in a silica container for a self absorption correction. All measurements were taken at ambient temperature, except the sample measurement, where a vanadium furnace was used to heat the sample to 800 °C. The essential diffraction parameters are shown in Table 7.1. Significant care was taken in minimising the background to avoid the possibility of an Al peak coincident with the InSe first peak [13]

7.3 Results

Figure 7.3 shows the $F(q)$ data for the three different wavelengths. In each case the data has had the background (corrected for sample masking effects) removed, the corrected container removed, has been normalised to vanadium, and has been corrected for multiple scattering, attenuation and inelastic scattering, as discussed in Sect. 3.2.7.3. From this it is immediately obvious that the statistical errors become significantly worse as wavelength decreases. This is due to the combination of

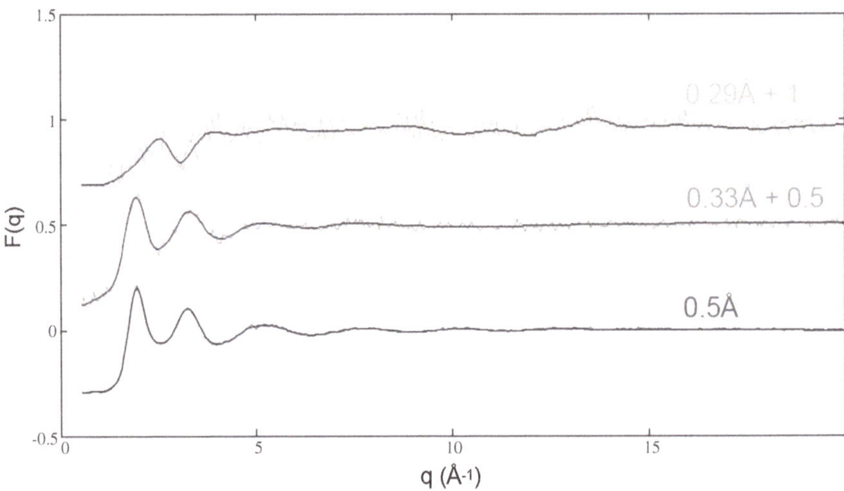

Fig. 7.3 The total structure factors for InSe at 0.5 Å (*red*), 0.33 Å (*green*) and 0.29 Å (*light blue*). The *black lines* through the data are spline fits

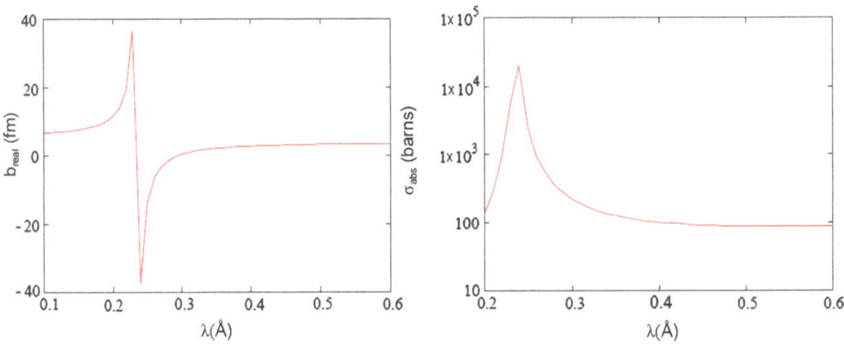

Fig. 7.4 The variation in the real scattering length and the absorption cross section with λ due to the ^{115}In absorption resonance

considerably reduced flux (~10x less at 0.33 Å than 0.5 Å) and the increased σ_{abs} caused by the variation of ^{115}In b_{imag} (Fig. 7.4). In order to reduce the effects of the statistical error the $F(q)$'s were spline fitted prior to Fourier transforming to real space. Although in the case of the 0.5 and 0.33 Å data this resulted in reasonable total pair correlation functions (Fig. 7.5), the 0.29 Å $F(q)$ suffered too significantly from statistical error. It should be highlighted that no attempt at a resolution correction [9, 23] was made.

As previously discussed, a resonance at thermal neutron energies leads to a variation in both the real and imaginary parts of the scattering length. This was calculated using (3.20) and (3.21) with the parameters taken from [19] (Table 7.2).

Table 7.2 Variation in incoherent scattering cross section, σ_{inc}, with wavelength due to the ^{115}In absorption resonance

λ (Å)	σ_{inc} for In-Nat (fm)
0.5	0.93
0.33	2.24
0.29	4.51

These variations lead to different weighting factors at the three wavelengths, as shown in Fig. 7.6. As ^{115}In has a non-zero spin it was important to determine that it is the b_+ scattering length [12] that experiences the resonance at 1.457 eV. As $b_+ = 2.1$ fm and $b_- = 6.4$ fm at 1.798 Å, the decrease in b_+ as the resonance is approached leads to an increase in the spin contribution to the incoherent scattering (Table 7.2), as can be seen from:

$$b_{inc}^2 = g_+ g_- (b_+ - b_-)^2 \tag{7.3}$$

where for ^{115}In $g_+ = 0.55$ and $g_- = 0.45$.

7.4 Discussion

Due to the complicated analysis procedure it is important to consider the veracity of the results. The low r cutoffs and their corresponding back Fourier transforms are shown in Fig. 7.5. It is clear that there is a significant difference in the total structure factor and the back Fourier transform in the case of the 0.33 Å data. It is probable that this difference is related to the difficulties in calculating the attenuation; due to the absorption resonance, considerations which are usually insignificant, such as the beam profile or the wavelength resolution, become significant. Combined with this is the possibility of bubbles forming in liquid which would cause a decrease in the absorption that is difficult to estimate. A similar difference, but of a much reduced magnitude, are also seen in the 0.5 Å data. It should also be noted that there is a slight (~ 0.1 Å) broadening of the first peak in the 0.33 Å data, which is probably related to the decreased resolution at this wavelength.

In back Fourier transforming the 0.5 Å data, the low q region only became consistent when the number density was reduced from 0.0312 (Table 7.1) to 0.03. This number density compares well with the number density of 0.0299 calculated from extrapolating between the two endpoints liquid In [17] and Se [13].

Figure 7.5a compares the Lorch modified 0.5 Å total pair correlation function with that determined from neutron scattering by Lague [13] taken on D4b (also at 0.5 Å), and from the EXAFS modified ab initio MD of Ferlat et al. [4]. Although there is a slight difference in the peak heights when compared with the Lague data, the peak positions are in good agreement (Table 7.3). Despite the difference in peak heights the first peak coordination numbers from both this work (Table 7.4) and that of Lague [13] are ~3, which is the In–Se coordination number in the corresponding crystal. However when comparing (Fig. 7.7) with the AIMD data of Ferlat et al. [4], the height of the first peak, which is due mainly to the heteropolar

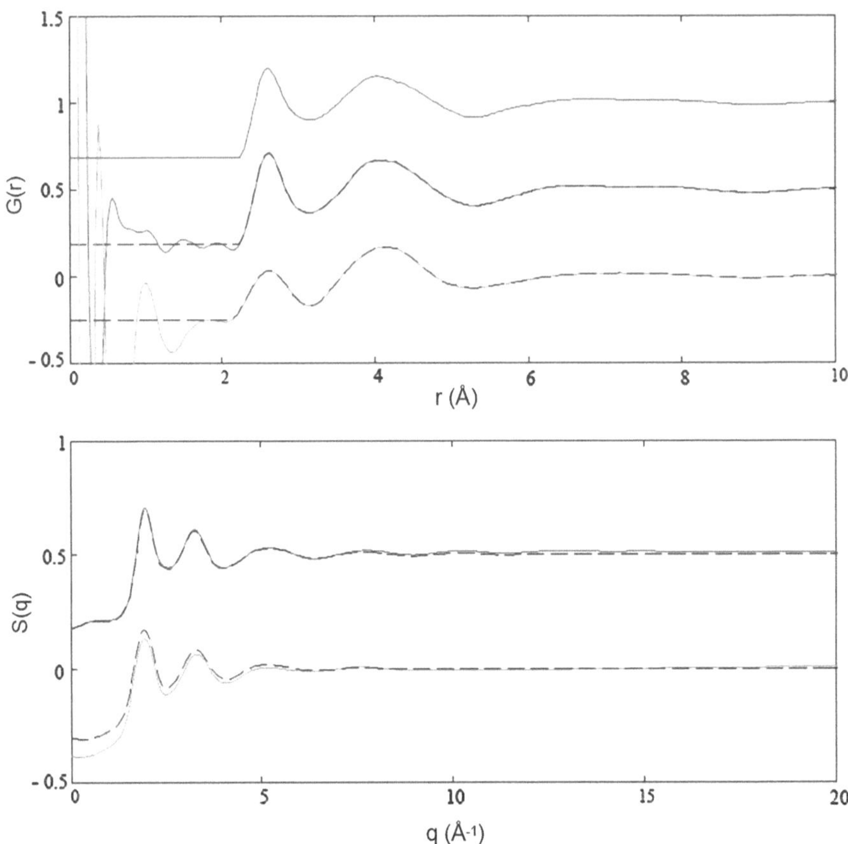

Fig. 7.5 **a** The Lorch modified $G(r)$'s for 0.5 Å (*red*) and 0.33 Å (*green*) compared with neutron diffraction data of Lague [13] also taken at 0.5 Å. The *broken black lines* show the low r cutoff data. There is a small broadening (~0.1 Å) of the first peak in the 0.33 Å data which is probably related to the diffractometer resolution function [23]. **b** The back Fourier transforms (*broken black lines*) of the cutoff data shown in (**a**) compared with the original $F(q)$'s (*solid coloured lines*). There is a significant difference in the case of the 0.33 Å data that is probably due to an incorrect absorption correction

Table 7.3 A comparison of the first peak positions between both wavelengths and the data of Lague [13]

Wavelength (Å)	First peak position (Å)
0.5	2.62 ± 0.03
0.33	2.62 ± 0.03
0.5 [13]	2.60 ± 0.02

bonds, shows a distinct difference. Similarly in the 0.33 Å data the second peak is in good agreement while the first peak height is significantly different. When comparing the coordination numbers this peak height difference is evident, as the coordination number from the AIMD of Ferlat et al. [4] is 4.2.

Table 7.4 The In–Se coordination numbers determined from both wavelengths. The running coordination numbers were calculated based up to an r value where $G(r)$ was at a minimum (3.15 Å)

Wavelength (Å)	Running	Gaussian	Reflected peak
0.5	2.9 ± 0.1	2.9 ± 0.1	2.2 ± 0.1
0.33	3.0 ± 0.2	3.0 ± 0.2	2.5 ± 0.2

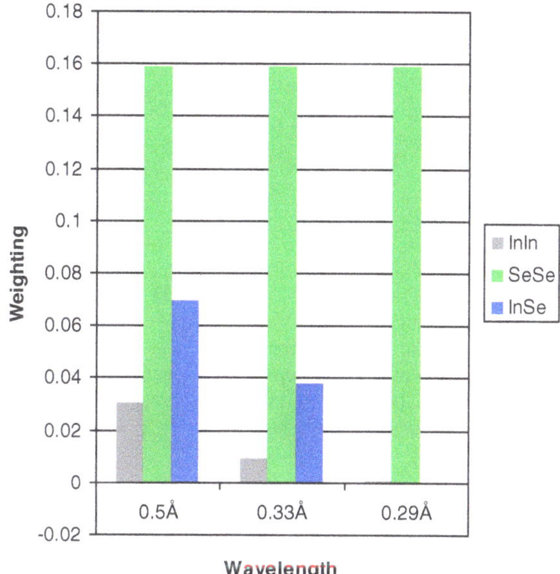

Fig. 7.6 The calculated relative weighting factors of the In–In, Se–Se and In–Se partial pairs at the three experimental wavelengths. As the Se has no wavelength dependence, the Se–Se weighting is constant

Despite the absorption correction problems with the 0.33 Å, its peak position and coordination number are comparable to that of the 0.5 Å data; both datasets suggest that there is no significant change in the In–Se correlations upon melting. This is a significant result as each dataset has a different In–Se weighting (Fig. 7.6). If this is the case it appears there is minimal In–In homopolar bonding contributing to the first peak. This suggests that homopolar bonding is not responsible for the low conductivity behaviour in InSe as posited by Ferlat et al. [4]. However due to the increased uncertainty associated with neutron diffraction from a highly absorbing resonant sample, it is important that these conclusions are confirmed; X-ray diffraction measurements could be used for this as the scattering intensities would effectively be inverted. These results could then be combined to rigorously constrain a MD-RMC.

When considering whether or not the ANS method can be applied to levitation samples, we need to consider the payoff between smaller samples giving less flux, and the reduction in absorption due both to reduced sample size and spherical geometry. For example in the case of InSe, a 3 mm diameter hemispherical

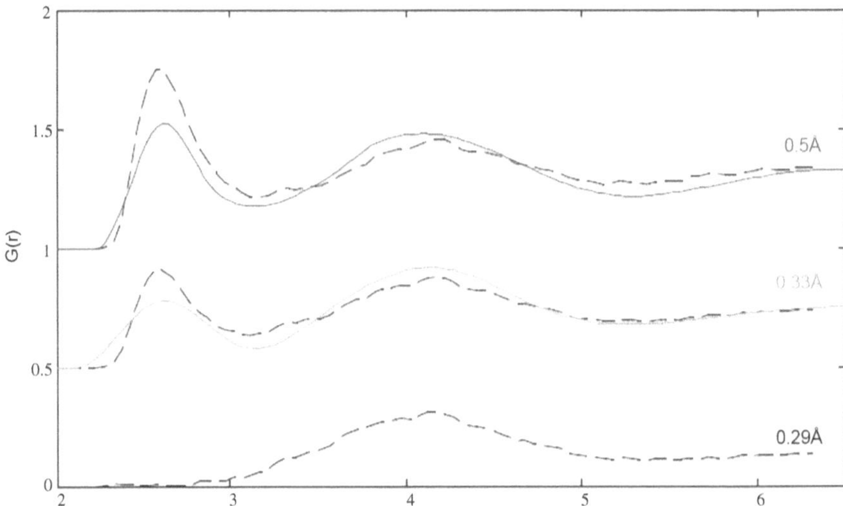

Fig. 7.7 The $G(r)$'s for 0.5 Å (*red*) and 0.33 Å (*green*) data compared with the weighted sums of the partial pair correlation functions of Ferlat et al. [4] (*broken black lines*). The weighted sum for 0.29 Å is shown for comparison. It is important to note that no attempt to convolute the weighted partial pair correlation functions with an experimental resolution has been made. While the second peak height is in reasonable agreement, there is a significant difference in the first peak height for both wavelengths

spherical sample at 0.33 Å would result in the transmission doubling. However this would also decrease of a factor of ~100 in the flux from the sample, which would result in impractically long counting times given the small sample mass loss during levitation. Due to this it is probable that the combination of ANS and aerodynamic levitation is only feasible for samples where the resonant element is dilute; an example would be the liquid precursors of rare earth doped glasses [25, 26].

7.5 Conclusions

The atomic structure in the liquid InSe was investigated with anomalous neutron diffraction, in order to determine the mechanism for its unusual semiconducting behaviour [21]. Measurements were taken at 0.33 and 0.5 Å, attempting to take advantage of the absorption resonances present in ^{115}In. Due to difficulties correcting the 0.33 Å data for attenuation a first order difference determination was not possible; however the coordination number of ~3 determined at both wavelength, which is the same as the crystalline structure, implies that there is minimal In–In homopolar contributing to the total pair correlation function first peak.

References

1. Anderson PW (1958) Absence of diffusion in certain random lattices. Phys Rev 109(1):1492–1505
2. Barnes AC, Guo C (1994) The structure of liquid thallium selenide. J Phys: Condens Matter 6(23A):A229–A234
3. Enderby JE, Barnes AC (1990) Liquid semiconductors. Rep Prog Phys 53(2):85–179
4. Ferlat G, San Miguel A, Xu H, Aouizerat A, Blase X, Zuniga J, Munoz-Sanjose V (2004) Semiconductor-metal transitions in liquid $In_{100-x}Se_x$ alloys: a concentration-induced transition. Phys Rev B 69(15):1–10
5. Ferlat G, Martínez-García D, San Miguel A, Aouizerat A, Muñoz-Sanjosé V (2004) High pressure–high temperature phase diagram of InSe. High Press Res 24(1):111–116
6. Fischer HE, Cuello GJ, Palleau P, Feltin D, Barnes AC, Badyal YS, Simonson JM (2002) D4c: a very high precision diffractometer for disordered materials. Appl Phys A 74:S160–S162
7. Greenwood DA (1958) The Boltzmann equation in the theory of electrical conduction in metals. Proc Phys Soc Lond 71(460):585–596
8. Holender J, Gillan M (1996) Composition dependence of the structure and electronic properties of liquid Ga-Se alloys studied by ab initio molecular dynamics simulation. Phys Rev B: Condens Matter 53(8):4399–4407
9. Howells WS (1984) A deconvolution procedure for liquid structure determination. Nucl Instrum Methods Phys Res 219(3):543–552
10. Hudgens S, Johnson B (2011) Overview of phase-change chalcogenide nonvolatile memory technology. MRS Bull 29(11):829–832
11. Kittel C (2005) Introduction to solid state physics, 8th edn. Wiley, Hoboken
12. Koester L, Knopf K (1980) Slow neutron scattering lengths of the isotopes of silver and indium. Z Phys A: Hadrons Nucl 297(1):85–91
13. Lague SB (1996) The structural and electronic properties of some liquid semiconductors. Ph. D., University of Bristol
14. Lague SB, Barnes AC, Archer AD, Howells WS (1996) The electronic properties and structure of liquid Tl-Se and Ga-Se alloys. J Non-Cryst Solids 205–207:89–93
15. Lee H, Kim YK, Kim D, Kang D-H (2005) Switching behavior of indium selenide-based phase-change memory cell. IEEE Trans Magn 41(2):1034–1036
16. Likforman A, Carré D, Etienne J, Bachet B (1975) Crystal-structure of indium monoselenide (InSe). Acta Crystallogr Sect B 31(May15):1252–1254
17. McGonigal PJ, Cahill JA, Kirshenbaum AD (1962) The liquid range density, observed normal boiling point and estimated critical constants of indium. J Inorg Nucl Chem 24:1012–1013
18. Mott NF, Davis EA (1971) Electronic processes in non-crystalline materials. Clarendon Press, Oxford
19. Mughabhab SF (2006) Altas of Neutron Resonances. 5th edn. Elsevier, Amsterdam
20. Ohno S, Barnes AC, Enderby JE (1994) The electronic properties of liquid $Ag_{1-x}Se_x$. J Phys: Condens Matter 6(28):5335–5350
21. Okada T, Ohno S (1993) Electrical properties of liquid In–Se alloys. J Non-Cryst Solids 156–158:748–751
22. Okada T, Ohno S (1997) Electrical properties of liquid Ga-Se and In–Se alloys. J Phys Soc Jpn 66(10):3171–3177
23. Salmon PS, Petri I, de Jong PHK, Verkerk P, Fischer HE, Howells WS (2004) Structure of liquid lithium. J Phys: Condens Matter 16(3):195–222
24. Strehlow WH, Cook EL (1973) Compilation of energy band gaps in elemental and binary compound semiconductors and insulators. J Phys Chem 2(1):163–199
25. Wright AC, Etherington G, Erwin Desa JA, Sinclair RN (1982) Neutron diffraction studies of rare earth ions in glasses. J de Physique 43(NC-9):31–34

26. Wright AC, Cole JM, Newport RJ, Fisher CE, Clarke SJ, Sinclair RN, Fischer HE, Cuello GJ (2007) The neutron diffraction anomalous dispersion technique and its application to vitreous $Sm_2O_3{\cdot}4P_2O_5$. Nucl Instrum Methods Phys Res Sect A 571(3):622–635
27. Ziman JM (1979) Models of disorder. Cambridge Press, Cambridge

Chapter 8
Conclusions

In this thesis I have shown how levitation techniques can be combined with neutron diffraction, XAS, and computer simulations, in order to determine the structures of a diverse range of amorphous materials. The structural information reported relates to several distinct areas of condensed matter; the novel glass systems of $BaTi_2O_5$ (BTO) and rare earth doped BTO; the controversial issue of isocompositional first order liquid-liquid phase transitions and polyamorphism in yttria aluminate liquids; and the extension of Invar into the liquid state to provide further constraints on crystalline models.

In the case of BTO glass and the four (Nd, Gd, Er, Yb) 3.5 % rare earth doped BTO glasses, neutron diffraction data at the total structure factor level were presented. These total structure factors were used as the input to RMC refinements of a pair potential MD simulated starting configuration. The results of both the neutron diffraction data and the RMC were compared with XAS taken at the Ti K-edge and RE L_3-edge. The neutron diffraction data established Ti–O coordination numbers of between 4 and 5, which was consistent with the limited XANES analysis employed. The RMC determined the underlying Ti–O network to consist of a distribution of 4, 5, and to a smaller extent 6 coordinated polyhedra. Introduction of the rare earth did not appear to significantly alter the network structure, and the RE-O bond length decreased with decreasing rare earth ionic radius. The extent to which the rare earth atoms formed clustered was demonstrated to be minimal, which suggests that BTO glasses may be good candidates as hosts for fluorescent rare earth atoms. In order to develop this, further structural experiments should be combined with measurements of the optical properties.

In situ small angle neutron scattering measurements (SANS) of three $(Y_2O_3)_x$ $(Al_2O_3)_{1-x}$ liquids ($x = 0.2, 0.25, 0.375$) did not exhibit a rise in intensity associated with an increase in density fluctuations due to the formation of a second liquid phase. These measurements were taken over a wide range of temperatures, including the purported [2] liquid–liquid phase transition temperature for $x = 0.2$ (1,788 K), which was determined by Greaves et al. [2] due to a significant rise in the small angle X-ray scattering intensity. The experimental difficulties present in the SANS

T. Farmer, *Structural Studies of Liquids and Glasses Using Aerodynamic Levitation*, Springer Theses, DOI: 10.1007/978-3-319-06575-5_8,
© Springer International Publishing Switzerland 2015

measurements, especially the occurrence of surface scattering, were carefully considered and established to be sufficiently small that a liquid–liquid phase transition would be clearly observable. It was suggested that the difficulty in accurately determining the X-ray transmission might account for the difference in the results. The apparent observation by Greaves et al. [2] of a 'polyamorphic rotor' was investigated by pyrometric studies coupled with video of the sample behaviour. In the case of sample behaviour identical to that of Greaves et al. [2], it was established to be related to the movement of a small (~0.15 mm diameter) low density region was responsible for the pyrometric oscillations. As this behaviour was demonstrated in different compositions, at multiple temperatures, and was dependent on only a very small region (in contrast to the third of the sample suggested by Greaves et al. [2] it is incompatible with the concept of a 'polyamorphic rotor'. From these results no evidence of a liquid–liquid phase transition was observed.

In situ aerodynamic levitation and neutron diffraction with isotopic substitution was performed on $Fe_{65}Ni_{35}$ (Invar) using ^{Nat}Ni, ^{58}Ni, and ^{60}Ni, and also on pure Fe. The coordination number of monatomic iron was determined to be ~12, in agreement with that determined by Holland-Moritz et al. [3], although the first peak position was found to be 0.05 Å lower at 2.50 Å. The similarity between the pure iron structure factor and the total structure factors of the three Invar samples suggested that the Ni effectively substituted for Fe atoms without changing the structure of pure iron. This was also suggested by the first peak position in the total correlation function which was at 2.49 Å. However, analysis of the Bhatia-Thornton partial structure factors indicated a small degree of chemical ordering, showing that the Ni substitution cannot be considered as a completely random substitution. The small angle region of the structure factors of the Invar samples showed a significant rise corresponding to magnetic correlations within the liquid, similar in magnitude to pure iron. The fact that relatively strong correlations are present in the Invar samples at ~1,750 K, despite its relatively small Curie temperature [1] (~500 K), is unusual when compared to the weak correlations seen in Ni [4].

Alongside these levitation based structural measurements was an anomalous neutron scattering (ANS) experiment on liquid InSe. This had the dual purpose of investigating the anomalous liquid semiconducting behaviour of InSe, and determining the feasibility of combining ANS with in situ aerodynamic levitation. The results show it is difficult to correct for the extreme attenuation which meant that a first order difference determination was not possible. The fact that both the 0.5 and 0.33 Å data determined a first peak coordination number of ~3 (the same as crystalline InSe) suggests that it is unlikely that there is significant In–In homopolar bonding at distances <3.15 Å. Further experimental measurements such as X-ray diffraction combined with simulation techniques may help to confirm this. The challenging experimental conditions resulting from an absorption resonance appear incompatible with in situ aerodynamic levitation. The exception to this might occur in samples with a very dilute amount of the resonant element.

References

1. Crangle J, Hallam GC (1963) The magnetization of face-centred cubic and body-centred cubic iron + nickel alloys. Proc R Soc A: Math Phys Eng Sci 272(1348):119–132
2. Greaves GN, Wilding MC, Fearn S, Langstaff D, Kargl F, Cox S, Van QVu, Majérus O, Benmore CJ, Weber R, Martin CM, Hennet L (2008) Detection of first-order liquid/liquid phase transitions in yttrium oxide-aluminum oxide melts. Science 322:566–570
3. Holland-Moritz D, Schenk T, Convert P, Hansen T, Herlach DM (2005) Electromagnetic levitation apparatus for diffraction investigations on the short-range order of undercooled metallic melts. Meas Sci Technol 16(2):372–380
4. Weber M, Knoll W, Steeb S (1978) Magnetic small-angle scattering from molten elements iron, colbalt and nickel. J Appl Crystallogr 11:638–641